ANIMALS IN TRANSLATION

Also by Temple Grandin

Emergence: Labeled Autistic *(with Margaret M. Scariano)*

Thinking in Pictures: And Other Reports from My Life with Autism

Genetics and the Behavior of Domestic Animals

Livestock Handling and Transport

Also by Catherine Johnson

Shadow Syndromes: The Mild Forms of Major Mental Disorders
That Sabotage Us *(with John J. Ratey)*

Lucky in Love: The Secrets of Happy Couples
and How Their Marriages Thrive

When to Say Goodbye to Your Therapist

ANIMALS IN TRANSLATION

The Woman Who Thinks Like A Cow

**Temple Grandin
and Catherine Johnson**

B L O O M S B U R Y
LONDON • NEW DELHI • NEW YORK • SYDNEY

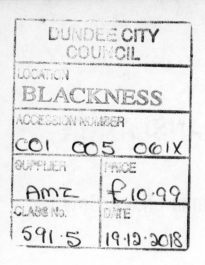

DUNDEE CITY
COUNCIL

LOCATION
BLACKNESS

ACCESSION NUMBER
CO1 005 061X

SUPPLIER | PRICE
AMZ | £10.99

CLASS No. | DATE
591·5 | 19.12.2018

First published in Great Britain 2005
This paperback edition published 2006

Copyright © 2005 by Temple Grandin and Catherine Johnson

The moral right of the author has been asserted

Bloomsbury Publishing Plc, 50 Bedford Square, London WC1B 3DP

Bloomsbury Publishing, London, New Delhi, New York and Sydney

A CIP catalogue record for this book is available from the British Library

ISBN 978 0 7475 6669 4

14

MIX
Paper from
responsible sources
FSC® C013604

Printed in Great Britain by CPI Group (UK) Ltd, Croydon CR0 4YY

www.bloomsbury.com/templegrandin

For the animals
—*Temple Grandin*

For Jimmy, Andrew, and Christopher
—*Catherine Johnson*

Contents

Animals
in
Translation

1. My Story

People who aren't autistic always ask me about the moment I realized I could understand the way animals think. They think I must have had an epiphany.

But it wasn't like that. It took me a long time to figure out that I see things about animals other people don't. And it wasn't until I was in my forties that I finally realized I had one big advantage over the feedlot owners who were hiring me to manage their animals: being autistic. Autism made school and social life hard, but it made animals easy.

I had no idea I had a special connection to animals when I was little. I liked animals, but I had enough problems just trying to figure out things like why a really small dog isn't a cat. That was a big crisis in my life. All the dogs I knew were pretty big, and I used to sort them by size. Then the neighbors bought a dachshund, and I was totally confused. I kept saying, "How can it be a dog?" I studied and studied that dachshund, trying to figure it out. Finally I realized that the dachshund had the same kind of nose my golden retriever did, and I got it. Dogs have dog noses.

That was pretty much the extent of my expertise when I was five.

I started to fall in love with animals in high school when my mother sent me to a special boarding school for gifted children with emotional problems. Back then they called everything "emotional problems." Mother had to find a place for me because I got kicked out of high school for fighting. I got in fights because kids teased me. They'd call me names, like "Retard," or "Tape recorder."

They called me Tape Recorder because I'd stored up a lot of phrases in my memory and I used them over and over again in every conversation. Plus there were only a few conversations I liked to have, so that amplified the effect. I especially liked to talk about the rotor ride at the carnival. I would go up to somebody and say, "I went to Nantasket Park and I went on the rotor and I really liked the way it pushed me up against the wall." Then I would say stuff like, "How did you like it?" and they'd say how they liked it, and then I'd tell the story all over again, start to finish. It was like a loop inside my head, it just ran over and over again. So the kids called me Tape Recorder.

Teasing hurts. The kids would tease me, so I'd get mad and smack 'em. That simple. They always started it, they liked to see me react.

My new school solved that problem. The school had a stable and horses for the kids to ride, and the teachers took away horseback riding privileges if I smacked somebody. After I lost privileges enough times I learned just to cry when somebody did something bad to me. I'd cry, and that would take away the aggression. I still cry when people are mean to me.

Nothing ever happened to the kids who were teasing.

The funny thing about the school was, the horses had emotional problems, too. They had emotional problems because in order to save money the headmaster was buying cheap horses. They'd been marked down because they had gigantic behavior problems. They were pretty, their legs were fine, but emotionally they were a mess. The school had nine horses altogether, and two of them couldn't be ridden at all. Half the horses in that barn had serious psychological problems. But I didn't understand that as a fourteen-year-old.

So there we all were up at boarding school, a bunch of emotionally disturbed teenagers living with a bunch of emotionally disturbed animals. There was one horse, Lady, who was a good horse when you rode her in the ring, but on the trail she would go berserk. She would rear, and constantly jump around and prance; you had to hold her back with the bridle or she'd bolt to the barn.

Then there was Beauty. You could ride Beauty, but he had very nasty habits like kicking and biting while you were in the saddle. He would swing his foot up and kick you in the leg or foot, or turn his head around and bite your knee. You had to watch out. Whenever you tried to mount Beauty he kicked *and* bit—you had both ends coming at you at the same time.

But that was nothing compared to Goldie, who reared and plunged whenever anyone tried to sit on her back. There was no way to ride that horse; it was all you could do just to stay in the saddle. If you did ride her, Goldie would work herself up into an absolute sweat. In five minutes she'd be drenched, dripping wet. It was flop sweat. Pure fear. She was terrified of being ridden.

Goldie was a beautiful horse, though; light brown with a golden mane and tail. She was built like an Arab horse, slender and fine, and had perfect ground manners. You could walk her on a lead, you could groom her, you could do anything you liked and she was perfectly behaved just so long as you didn't try to ride her. That sounds like an obvious problem for any nervous horse to have, but it can go the other way, too. I've known horses where people say, "Yeah you can ride them, but that's all you can do with them." That kind of horse is fine with people in the saddle, and nasty to people on the ground.

All the horses at the school had been abused. The lady they bought Goldie from had used a nasty, sharp bit and jerked on it as hard as she could, so Goldie's tongue was all twisted and deformed. Beauty had been kept locked in a dairy stanchion all day long. I don't know why. These were badly abused animals; they were very, very messed up.

But I had no understanding of this as a girl. I was never mean to the horses at the school (other kids were sometimes), but I wasn't any horse-whispering autistic savant, either. I just loved the horses.

I was so wrapped up in them that I spent every spare moment working the barns. I was dedicated to keeping the barn clean, making sure the horses were groomed. One of the high points of my high school career was the day my mom bought me a really nice English bridle and saddle. That was a huge event in my life, because it was mine, but also because the saddles at school were so crummy. We rode on old McClellands, which were honest-to-god cavalry saddles first used in the Civil War. The school's saddles probably went back to World War II when they still had some horse units in the army. The McClelland was designed with a slot down the center of it to spare the horse's back. The slot was good for the horse but horrible for the rider. I don't think there's ever been a more uncomfortable saddle on earth, though I have to say that when I read about the Northern Alliance soldiers in Afghanistan riding on saddles made out of wood, that sounded worse.

Boy did I take care of that saddle. I loved it so much I didn't even leave it in the tack room where it belonged. I brought it up to my dorm room every day and kept it with me. I bought special saddle soap and leather conditioner from the saddle shop, and I spent hours washing and polishing it.

As happy as I was with the horses at school, my high school years were hard. When I reached adolescence I was hit by a tidal wave of anxiety that never stopped. It was the same level of anxiety I felt later on when I was defending my dissertation in front of my thesis committee, only I felt that way all day long and all night, too. Nothing bad happened to make me so anxious all of a sudden; I think it was just one of my autism genes kicking into high gear. Autism has a lot in common with obsessive-compulsive disorder, which is listed as an anxiety disorder in the *Diagnostic and Statistical Manual*.

Animals saved me. One summer when I was visiting my aunt, who had a dude ranch in Arizona, I saw a herd of cattle being put through the *squeeze chute* at a neighboring ranch. A squeeze chute is an apparatus vets use to hold cattle still for their shots by squeezing them so tight they can't move. The squeeze chute looks like a big V made out of metal bars hinged together at the bottom. When a cow walks into the chute an air compressor closes up the V, which squeezes the cow's body in place. The rancher has plenty of space for his hands and the hypodermic needle between the metal bars. You can find pictures of them on the Web if you want to see what they look like.

As soon as I caught sight of that thing I made my aunt stop the car so I could get out and watch. I was riveted by the sight of those big animals inside that squeezing machine. You might think cattle would get really scared when all of a sudden this big metal structure clamps together on their bodies, but it's exactly the opposite. They get really calm. When you think about it, it makes sense, because deep pressure is a calming sensation for just about everyone. That's one of the reasons a massage feels so good—it's the deep pressure. The squeeze chute probably gives cattle a feeling like the soothing sensation newborns have when they're swaddled, or scuba divers have underwater. They like it.

Watching those cattle calm down, I knew I needed a squeeze chute of my own. When I got back to school that fall, my high school teacher helped me build my own squeeze chute, the size of a human being down on all fours. I bought my own air compressor, and I used plywood boards for the V. It worked beautifully. Whenever I put myself inside my squeeze machine, I felt calmer. I still use it today.

I got through my teenage years thanks to my squeeze machine and my horses. Animals kept me going. I spent every waking minute that I didn't have to be studying or going to school with those horses. I even rode Lady at a show. It's hard to imagine today, a school keeping a stable of emotionally disturbed and dangerous horses for its underaged students to ride. These days you can't even play dodgeball in gym class because somebody might get hurt. But that's the way it was. A lot of us got nipped or stepped on or thrown at that school, but no one was ever seriously hurt, at least not while I was there. So it worked out.

I wish more kids could ride horses today. People and animals are supposed to be together. We spent quite a long time evolving together, and we used to be partners. Now people are cut off from animals unless they have a dog or a cat.

Horses are especially good for teenagers. I have a psychiatrist friend in Massachusetts who has a lot of teenage patients, and he has a whole different set of expectations for the ones who ride horses. He says that if you take two kids who have the same problem to the same degree of severity, and one of them rides a horse regularly and the other one doesn't, the rider will end up doing better than the nonrider. For one thing, a horse is a huge responsibility, so any teenage kid who's looking after a horse is developing good character. But for another, riding a horse isn't what it looks like: it isn't a person sitting in a saddle telling the horse what to do by yanking on the reins. Real riding is a lot like ballroom dancing or maybe figure skating in pairs. It's a relationship.

I remember looking down to make sure my horse was on the right lead. When a horse is cantering around the ring one of his front hooves has to thrust out farther forward than the other one, and the rider has to help him do that. If I leaned my body just the right way, it helped my horse get on the right lead. My sense of balance was so bad I could never learn to parallel ski no matter how hard I tried, though I did reach the advanced snowplow stage. Yet there I was, moving my body in sync with the horse's body to help him run right.

Horseback riding was joyous for me. I can remember being on a horse sometimes and we'd gallop in the pasture and that was such a big thrill. Of course it's not good for horses to run them all the time, but once in a while we'd get to have a little run, and I'd feel exhilarated. Or we'd be out on a trail riding, and do a really fast gallop down the road. I remember what it looked like, the trees whizzing by; I remember that really well to this day.

Riding becomes instinctual after a while; a good rider and his horse are a team. It's not a one-way relationship, either; it's not just the human relating to the horse and telling him what to do. Horses are super-sensitive to their riders and are constantly responding to the riders' needs even without being asked. School horses—the horses a stable uses to teach people how to ride—will actually stop trotting when they feel their rider start to lose his balance. That's why learning to ride a horse is completely different from learning to ride a bicycle. The horses make sure nobody gets hurt.

The love a teenager gets from a horse is good for him, and so is the teamwork. For years people always said you needed to send difficult kids to military school or the army. A lot of times that works because those places are so highly structured. But it would work a lot better if military schools still had horses.

———

Animals in Translation comes out of the forty years I've spent with animals.

It's different from any other book I've read about animals, mostly because I'm different from every other professional who works with animals. Autistic people can think the way animals think. Of course, we also think the way people think—we aren't *that* different from normal humans. Autism is a kind of way station on the road from animals to humans, which puts autistic people like me in a perfect position to translate "animal talk" into English. I can tell people why their animals are doing the things they do.

I think that's why I was able to become successful in spite of being autistic. Animal behavior was the right field for me, because what I was missing in social understanding I could make up for in understanding animals. Today I've published over three hundred scientific papers, my Web site gets five thousand visitors each month, and I give thirty-five lectures on animal management a year. I give another twenty-five or so on autism, so I'm on the road most of the time. Half the cattle in the United States and Canada are handled in humane slaughter systems I've designed.

I owe a lot of this to the fact that my brain works differently.

Autism has given me another perspective on animals most professionals don't have, although a lot of regular people do, which is that animals are smarter than we think. There are plenty of pet owners and animal lovers out there who'll tell you "little Fluffy can think," but animal researchers have mostly dismissed this kind of thing as wishful thinking.

But I've come to realize that the little old ladies are right. People who love animals, and who spend a lot of time with animals, often start to feel intuitively that there's more to animals than meets the eye. They just don't know what it is, or how to describe it.

I stumbled across the answer, or what I think is part of the answer, almost by accident. Because of my own problems, I've always followed neuroscientific research on the human brain as closely as I've followed my own field. I had to; I'm always looking for answers about how to manage my own life, not just animals' lives. Following both fields at the same time led me to see a connection between human intelligence and animal intelligence the animal sciences have missed.

The literature on autistic savants sparked my discovery. Autistic savants are people who can do things like tell you what day of the week you were born based on your birth date, or calculate in their heads whether your street address is a prime number or not. They usually have IQs in the mentally retarded range, though not always, yet they can *naturally* do things no normal human being can even be *taught* to do, no matter how hard he tries to learn or how much time he spends practicing.

Animals are like autistic savants. In fact, I'd go so far as to say that animals might actually *be* autistic savants. Animals have special talents normal people don't, the same way autistic people have special talents normal people don't; and at least some animals have special forms of genius normal people don't, the same way some autistic savants have special forms of genius. I think most of the time animal genius probably happens for the same reason autistic genius does: a difference in the brain autistic people share with animals.

The reason we've managed to live with animals all these years without noticing many of their special talents is simple: we can't see those talents. Normal people never have the special talents animals have, so normal people don't know what to look for. Normal people can stare straight at an animal doing something brilliant and have no idea what they're seeing. Animal genius is invisible to the naked eye.

I'm sure I don't know all the talents animals have, either, let alone all the things they could use their talents to do if we gave them the chance. But now that I've seen the connection between autistic savantry and animal genius at least I have an idea what I'm looking for: I'm looking for ways animals can use their amazing ability to *perceive things humans can't perceive,* and to *remember highly detailed information we can't remember,* to make life better for everyone, animals and people alike. Just off the top of my head, here's a thought: we have service dogs for the blind—how about service dogs for the middle-aged whose memories are going? I'm willing to bet that just about any dog can remember where you put your car keys better than you can if you're over forty, and probably if you're under forty, too.

Or how about service dogs who remember where your kids left the remote control? I bet a dog could do this if you gave him the training.

Of course, I don't know this for a fact. I could be wrong. But for me, predicting animal talents is getting to be a little like astronomers predicting the existence of a planet nobody can see based on their understanding of gravity. I'm starting to be able to accurately predict animal talents nobody can see based on what I know about autistic talent.

*   *   *

ANIMALS FROM THE OUTSIDE IN

By the time I got to college I knew I wanted to learn about animals.

That was the 1960s, and the whole field of psychology was B. F. Skinner and behaviorism. Dr. Skinner was so famous that just about every college kid in the country had a copy of *Beyond Freedom and Dignity* on his bookshelf. He taught that all you needed to study was behavior. You weren't supposed to speculate about what was inside a person's or an animal's head because you couldn't measure all the stuff inside the *black box*—intelligence, emotions, motives. The black box was off-limits; you couldn't talk about it. You could measure only behavior, therefore you could study only behavior.[1]

For the behaviorists this was no great loss, since, according to them, environment was the only thing that mattered.

Some animal behaviorists took this idea to the extreme by teaching that animals didn't even *have* emotions or intelligence. Animals only had behavior, which was *shaped* by rewards, punishments, and positive and negative reinforcements from the environment.

Rewards and *positive reinforcers* are the same thing: something good happens to you because of something you did. *Punishment* and *negative reinforcement* are opposites. Punishment is when something bad happens to you because of something you did; negative reinforcement is when something bad *stops* happening to you, or doesn't *start* happening to you in the first place, because of something you did. Punishment is bad, and negative reinforcement is good. Punishment makes you stop doing what you're doing, although a lot of behaviorists believe that punishing a bad behavior isn't as effective as rewarding a good behavior when it comes to getting an animal to do what you want him to do.

Negative reinforcement is the hardest to understand. Negative reinforcement isn't a punishment; it's a reward. But the reward is *negative* in the sense that something you don't like either stops or doesn't start in the first place. Say your four-year-old is screaming and crying and giving you a headache. Finally you lose your patience and blow up at him, and he's shocked into silence. That's negative reinforcement, because you've made the crying go away, which is what you wanted. Now you're probably more likely to blow up at

him the next time he starts a tantrum, because you've been negatively reinforced for blowing up at him during this tantrum.

Behaviorists thought these basic concepts explained everything
about animals, who were basically just stimulus-response machines.
It's probably hard for people to imagine the power this idea had
back then. It was almost a religion. To me—to lots of people—B. F.
Skinner was a god. He was the god of psychology.

It turned out he wasn't much of a god in person. I met B. F.
Skinner once. I was probably eighteen years old at the time. I'd written him a letter about my squeeze machine, and he'd written me
back saying what impressed him was my motivation. Which is kind
of funny when you think about it. Here was the god of behaviorism
talking about my internal motivation instead of my behavior. I guess
he was ahead of his time, since motivation is a hot topic in autism
research today.

After I got his letter I called up his office and asked if I could
come see him. I wanted to talk to him about some of the research I
had done.

His office called and invited me down to Harvard for a visit. It
was like going to see the Pope at the Vatican. Dr. Skinner was the
most famous professor in all of psychology; he'd been on the cover
of *Time* magazine.[2] I was very nervous just about walking up to see
him. I remember walking to William James Hall and looking up at
the building feeling like "This is the temple of Psychology."

But when I went into his office, it was a big letdown. He was just
a normal-looking man. I remember he had this plant wired up
around his office, growing all around the room. We were sitting
there talking, and he started asking really personal questions. I don't
remember what they were, because I almost never remember specific
words and sentences from conversations. That's because autistic
people think in pictures; we have almost no words running through
our heads at all. Just a stream of images. So I don't remember the
verbal details of the questions; I just remember that he asked them.

Then he tried to touch my legs. I was shocked. I wasn't in a sexy
dress, I was in a conservative dress, and that was the last thing I
expected. So I said, "You may look at them, but you may not touch
them." I do remember saying that.

We did get to talk about animals and behavior, though, and finally I said to him, "Dr. Skinner, if we could just learn how the brain works." That's the other part of the conversation I remember specifically.

He said, "We don't need to learn about the brain, we have operant conditioning."

I remember driving back to school going over this in my mind, and finally saying to myself, "I don't think I believe that."

I didn't believe it because I had problems that sure didn't seem to be coming from my environment. Also, I'd taken an animal ethology class at college—ethologists study animals in their natural environments—and Thomas Evans, the teacher, had taught us about animal instincts, which were hardwired behavior patterns the animal was born with. Instincts had nothing to do with the environment, they came with the animal.

Dr. Skinner changed his mind when he got old. My friend John Ratey, a psychiatrist at Harvard who wrote the books *Shadow Syndromes* (with my co-author on this book, Catherine Johnson) and *A User's Guide to the Brain,* told me a story about a lunch he had with Dr. Skinner near the end of his life.[3] While they were talking John asked him, "Don't you think it's time we got inside the black box?"

Dr. Skinner said, "Ever since my stroke I've thought so."

The brain is pretty powerful, and a person whose brain isn't working right knows just how powerful. Dr. Skinner had to learn the hard way. His stroke showed him not everything is controlled by the environment. But back in the 1970s, when I was getting started, behaviorism was the law.

I don't want to sound like the enemy of behaviorism, though, because I'm not. In one way behaviorists weren't that different from ethologists, because neither group looked inside the animal's head. Behaviorists looked at animals in laboratory environments; ethologists looked at animals in their natural environment. But both were looking at animals from the outside.

Behaviorists made a big mistake declaring the brain off-limits, but their focus on the environment was a huge step forward and is to this day. Until behaviorism came along, probably no one understood how important the environment is. People still don't. In the meatpacking industry, where I've worked for thirty years designing humane handling systems, a lot of plant owners don't think twice about their cattle's environment. If there's a problem with the herd, it doesn't even occur to them to look at the animals' surroundings to see what's going on. People want the equipment that I install, but they don't realize that *the equipment won't work if the environment is bad.*

In a plant, the environment means the physical environment, and it also means the way the employees handle the animals. If the animal handling is bad, no amount of top-notch, well-maintained equipment is going to work.

The *center-track restraining system* I designed, which has been installed in half of all the plants in North America, works only when you have good animal handling. My restraining system is a conveyor belt that goes under the animal's chest and belly. The animals straddle it lengthwise the same way they would straddle a sawhorse.

The reason plants have adopted my design is that animals are much more willing to walk onto it than they are the old V-shaped restraining systems, so it's a lot more efficient. That was the only thing wrong with the old restraining systems: the animals didn't like walking onto them. The *V-restrainers* work fine, and they don't hurt the animals, but they squeeze the animal's feet together, and animals don't like to walk into a space where they feel like there isn't enough space for their feet. My design innovation wasn't technological, it was behavioral. It works better because it respects the animal's behavior.

But the plants don't seem to realize that, so naturally they also don't realize that if they have poor handling of their animals my equipment won't work. They focus on the equipment.

The other thing I like about behaviorists is that a lot of the time they're natural-born optimists. In the beginning, behaviorists thought the laws of learning were simple and universal, and all creatures followed them. That's why B. F. Skinner thought laboratory rats were the only animals anybody needed to look at, because all animals and people learned the same way.

Dr. Skinner's whole concept of learning was *associationist,* which meant that positive associations (or rewards) increased behavior, and negative associations (or punishment) decreased behavior. If you wanted to teach a really complex behavior, all you had to do was break it down into its component parts and teach each little, tiny step separately, giving rewards along the way. That was called *task analysis*, and it was a huge help not only for animal training (though animal trainers had always done this to some extent), but also for anybody trying to teach children or adults with disabilities. I've seen behavioral books for parents that take all the different things a child or adult has to do during the day, like get up, get dressed, eat breakfast, and so on, and break each activity down into its component parts. A supposedly simple thing like getting your clothes on in the morning might involve twenty or thirty different steps or more, and a task analysis lists each one, and you teach each one separately.

Doing a task analysis isn't as easy as it sounds, because nonhandicapped people aren't really aware of the very small, separate movements that go into an action like tying your shoe or buttoning your shirt. Typical kids pick these things up pretty easily, so parents don't have to be especially skilled to teach them how to put their clothes on or tie their shoes. If you've ever tried to teach shirt buttoning to a person who has absolutely no clue how to do it, you soon realize that you don't really know how to do it, either—not in the sense of knowing the sequence of tiny, separate motions that go into successfully buttoning a button. You just do it.

The behaviorists' belief that any animal or person could learn just about anything if the rewards were right led Ivar Lovaas to his work with autistic children. In his most famous study he took a group of very young autistic kids and gave one half of the children intensive *behavior therapy* while the other half got much less intensive treatment. Behavior therapy just meant *classical operant conditioning*, having the kids go over and over the behaviors Dr. Lovaas wanted them to learn and giving them rewards whenever they got something right. He published results showing that half of the kids who got the intensive therapy became "indistinguishable" from normal kids.[4]

There've been years of controversy over whether Dr. Lovaas did or didn't cure anybody, but to me, the fact that he brought those kids so far there could be an argument about it is what matters. Behaviorism gave parents and teachers a reason to think that autistic people were capable of a lot more than anybody thought, and that was a good thing.

The other major contribution behaviorists made is that they were, and still are today, fantastically close observers of animal and human behavior. They could spot tiny changes in an animal's behavior quickly, and connect the changes to something in the environment. That's one of my own most important talents with animals.

So for all of its problems, behaviorism had a lot to offer, and still does. Besides, the animal ethologists had their blind spots, too. For instance, both the ethologists and the behaviorists were in total agreement that practically the worst thing anyone could possibly do was to *anthropomorphize* an animal. Ethologists and behaviorists probably had different reasons for being against anthropomorphism—Dr. Skinner thought it was just as bad to anthropomorphize a person as an animal—but whatever the reasons, they agreed. Anthropomorphizing an animal was *wrong*.

To a large degree they were right to stress this, because humans just naturally treat their pets as if they're four-legged people a lot of the time. Professional trainers are constantly telling people not to assume their pets think and feel the same way they do, but people keep on doing it anyway. The dog trainer John Ross even has a story in his book *Dog Talk* about the first time he realized *he* was being anthropomorphic, and he's a professional. He had an Irish setter named Jason who was a big "garbage dog," constantly getting into the garbage whenever Mr. Ross wasn't around. Mr. Ross figured Jason knew he was being bad because if there was a mess on the floor the dog would take off running the minute Mr. Ross got home. On days when he hadn't gotten into the garbage he didn't run, so Mr. Ross thought this meant Jason knew that strewing garbage clear across the kitchen was wrong, and ran away because he felt bad.

He found out differently when a more experienced trainer had him try an experiment. He told Mr. Ross to go get into the garbage *himself*, when Jason wasn't watching, and dump it out all over the floor. Then he was supposed to bring Jason into the kitchen and see what the dog did.

It turned out Jason did what he always did when there was

garbage on the floor—he took off running. He wasn't running away because he felt guilty, he was running away because he felt scared. For Jason, garbage on the floor meant trouble. If Mr. Ross had stuck to behaviorist principles and thought about Jason's environment instead of about his "psychology," he wouldn't have made this mistake.[5]

A friend of mine had the same experience with her two dogs, a one-year-old German shepherd and a three-month-old golden retriever. One day the puppy pooped in the living room, and later on when the older dog saw the poop she got so anxious she started to drool. If the older dog had made the poop herself and then stood there drooling, her owner probably would have thought the dog knew she'd done something bad. But since the other dog had made the poop, her owner realized that the whole category of poop-on-living-room-floor was just plain bad news, period.

Those stories are classic examples of why it's not a good idea to anthropomorphize an animal, but that's not all there is to it. In my student days, even though *everyone* was against anthropomorphizing animals, I still believed it was important to think about the animal's point of view. I remember there was a great animal psychologist out of New Zealand named Ron Kilgour (he was an ethologist) who wrote a lot about the problem of anthropomorphizing. One of his early papers told a story about a person who had a pet lion he was shipping on an airplane. Someone thought the lion might like to have a pillow for the trip, the same way people do, so they gave him one, and the lion ate it and died. The point was: don't be anthropomorphic. It's dangerous to the animal.

But when I read this story I said to myself, "Well, no, he doesn't want a pillow, he wants something soft to lie on, like leaves and grass." I wasn't looking at the lion as a person, but as a lion. At least that's what I was trying to do.

That kind of thinking was illegal for behaviorists, however, and wasn't really encouraged by the ethologists, either. Both groups were environmentalists when you came right down to it, the big difference being which environment the animal was in while the researchers were studying him.

In the end, I had a pretty good grounding in animal ethology from undergraduate college before I started graduate school at Arizona State University. It was a good thing I did, because Arizona State was a hotbed of behaviorism. *Everything* was behaviorism. And I did not like some of the very cruel experiments they did to mice, rats, and monkeys. I remember one poor little monkey that had a little Plexiglas thing shoved onto his scrotum that they were shocking him with. I thought that was terrible.

I was not involved in any of the nasty experiments. I don't endorse using animals as subjects in experiments unless you're going to learn something incredibly important. If you're using animals to find a cure for cancer, that's different, especially since animals need a cure for cancer, too. But that's not what they were doing at Arizona. I spent one year in the psych department studying experimental psychology, and I thought, "I don't want to do this."

Even if the experiments had been fun for the animals, I still didn't see the point. My question was, "What are you learning from this?" Dr. Skinner wrote a lot about schedules of reinforcement, which is how often and how consistently the animal receives a reward for a particular behavior, and they were running every different schedule of reinforcement they could think of. Variable reinforcement, intermittent reinforcement, delayed reinforcement; you name it, they were running it.

It was totally artificial. What animals do in labs is nothing like what they do in the wild—so what are you actually learning when you do these experiments? You're learning how animals behave in labs. Finally people started doing things like letting a bunch of lab rats out in a courtyard and watching what they did. Suddenly the rats started developing complex behaviors no one had ever seen before.

SEEING THE WAY ANIMALS SEE: THE VISUAL ENVIRONMENT

The only research I was interested in doing at Arizona State was studying visual illusions in animals. I'm sure I was interested in visual illusions because I'm a visual thinker. I didn't know it at the time, but being a visual thinker was the start of my career with animals. It gave me an important perspective other students and professors didn't have, because animals are visual creatures, too. Animals are controlled by what they see.

When I say I'm a visual thinker I don't mean just that I'm good at making architectural drawings and designs, or that I can design my cattle-restraining systems in my head. I actually think in pictures. During my *thinking* process I have no words in my head at all, just pictures.

That's true no matter what subject I'm thinking about. For instance, if you say the word "macroeconomics" to me I get a picture of those macramé flowerpot holders people used to hang from their ceilings. That's why I can't understand economics or algebra; I can't picture it accurately in my mind. I flunked algebra. But other times thinking in pictures is an advantage. During the 1990s I knew all the dot-coms would go to hell, because when I thought about them the only images I saw were rented office space and computers that would be obsolete in two years. There wasn't anything real I could picture; the companies had no hard assets. My stockbroker asked me how I knew the two stock market crashes would happen, and I told him, "When the Monopoly play money starts jerking around the real money you're in trouble."

If I'm thinking about a structure I'm working on, all of my judgments and decisions about it happen in pictures. I see images of my design going together smoothly, images of problems and sticking points, or images of the whole thing collapsing if there's a major design flaw.

That's the point where words come in, *after* I've finished thinking it through. Then I'll say something like, "That won't work because it will collapse." My final judgment comes out in words, but not the process that led up to the judgment. If you think about a judge and jury, all my deliberations are in pictures, and only my final verdict is in words.

If I'm alone I'll say the verdict out loud, though I don't do it with other people around because I know I'm not supposed to. In college I did a lot of talking out loud because it helped me organize my thinking. A lot of autistic people talk out loud for the same reason. I'll also do some extremely simple running commentary in words. I'll say, "Let's try this," or, "Oh boy! I figured it out." The language is always simple. It's the pictures that are complex.

When I talk to other people I translate my pictures into stock phrases or sentences I have "on tape" inside my head. Those kids who called me Tape Recorder were right about me. They were mean, but they were right. I *am* a tape recorder. That's how I'm able to talk. The reason I don't sound like a tape recorder anymore is that I have so many stock phrases and sentences I can move around into new combinations. All my public speaking has been a huge help. When I got criticisms saying I always gave the same speech, I started moving my slides around. That moved my phrases around, too.

When I was young I had no idea that being a visual thinker made me different from anyone else. I thought everyone saw pictures inside their heads. So naturally, when I didn't like the lab work I was doing and wanted to start learning about animals in their natural environments, I focused on the visual environment. It wasn't a conscious decision, it was just what I naturally gravitated to.

Being verbal thinkers, behaviorists hadn't really thought about the visual environment. When they talked about the environment rewarding or punishing an animal in response to something it did, they usually meant food and electric shocks. That made sense for a Skinner box, where there's nothing much to look at, and if you mess up you get a shock. (A Skinner box was a special cage, usually a Plexiglas box, behaviorists used to test and analyze a rat's behavior. There was nothing in it except a lever and maybe some indicator lights that went on or off when a reward was available.) Most Skinner boxes didn't shock the animals, but if punishment was part of the experiment, usually the punishment would be a shock.

In the wild, though, there aren't any electric shocks, and you can't get food by pecking a lever. *You get food by being highly attuned to the visual environment.* Behaviorists finally started to catch on to the importance of vision to an animal when somebody did a famous experiment showing you could teach a monkey how to push a lever just by letting him look outside a window every time he hit the lever. They didn't need to give the monkey a food reward, just a view. Animals *need* to see, and they *want* to see.

While I was doing my research on visual illusions in the lab I started to hang out in feed yards with the cattle, where I noticed that a lot of times the animals didn't want to go through the chutes, which are the narrow passageways the cattle go through on the way to the squeeze chute. When I saw cattle balking and acting scared I just naturally thought, "Well let's look at it from the animal's point of view. I've got to get in the chute and see what he's seeing."

So I took pictures inside the chutes from the cattle's point of view. I even put black-and-white film in my camera because we thought animals saw in black and white. (Later on we learned that they see colors, too, but not in as wide a spectrum as we do.) I wanted to see what *they* were seeing.

That's when I noticed that simple things, like shadows or chains hanging down, made the animals balk.

The people at the feed yards thought my whole project was ridiculous. They couldn't imagine why I'd get in there and try to see what the cattle were seeing. Now I realize that in my own way I was being just as anthropomorphic as those people who gave the lion the pillow. Since I was a visual thinker I assumed cows were, too. The difference was I happened to be right.

When you're trying to understand how the environment is affecting an animal's behavior, you *have* to look at what the animal is seeing. I remember one time I went to a plant where they had a yellow metal ladder on a wall inside a building. The cattle had to go by it when they walked through a narrow alley. Those cattle just would not walk by that ladder. They'd plant their feet on the ground and refuse to move. Finally one of the yard people figured out the problem. He painted the ladder gray, and everything was fine. I work with management and with the employees down on the floor or in the yard, and I've found that a lot of times the guys in the yard are better at understanding animals than management.

If a cow sees a yellow raincoat flapping on a fence, she's in a panic. But if you aren't a visual thinker, it can be hard to even *notice* that yellow raincoat flapping on the fence. It doesn't jump out at normal people the way it does at me or at a cow.

Since I didn't realize other people thought in words instead of pictures, for a long time I could never figure out why so many animal handlers made such obvious, elementary mistakes. Not all of them do; I've met lots of good animal handlers in the meatpacking industry. But I was always surprised when I found an animal professional doing something that was just plain dumb. Why couldn't they see what they were doing wrong?

I remember one situation in particular, where the owner of a cattle-handling facility hired me as a last resort before they tore the whole place down and built it back up from the ground. He called me because his cattle wouldn't walk inside the narrow passage leading to the squeeze chute.

The problem wasn't that the cattle were afraid of getting their shots. Most cattle don't even know they're going to be getting shots inside the chute. Besides, a lot of animals barely feel their shots anyway. New dog owners are always surprised by this. They'll watch their dog cower and cringe as the vet examines him, then not blink an eye when he sticks him with a needle. Some vets say that's the difference between a dog, who isn't anticipating pain, and a person, who is. Thinking about a shot makes it worse.

The problem at the cattle-handling facility had to be something they were doing wrong, since those cattle were perfectly fine before they got there. But the owner couldn't figure it out. He needed to fix the situation fast, too, because skipping vaccinations isn't an option. Cattle aren't like children, who get vaccinated against a lot of diseases like polio or whooping cough that are pretty hard to catch nowadays. Cattle are extremely susceptible to bovine viral diarrhea and to respiratory diseases like pneumonia. If they don't get their shots, infectious disease will sweep through the herd and kill 10 percent of the animals. So you have to vaccinate, and in order to vaccinate you have to have your cattle walk into the squeeze chute. These cattle wouldn't do it, and the owner was starting to panic.

Things had gotten so bad the handlers were using cattle prods, which are fiberglass rods with two prongs on the end that deliver an electric shock to an animal. Prods will get an animal moving, but they're stupid things to use because they can panic the animals and make them rear up, which is dangerous for the workers. Prods always stress an animal, and when an animal is stressed his immune system goes down and he starts getting sick, which means higher veterinary bills. Plus stressed animals gain less weight, which means less meat to sell. Dairy cattle who've been handled with prods give less milk.

Stress is bad for human growth, too, although most people don't realize it. The one thing people do know about is *failure to thrive,* when children who've been badly abused or neglected suffer *stress dwarfism.* The child's biology is normal and he's eating enough food, but he doesn't grow. Stress dwarfism is pretty rare, but there's evidence that stressed children, just like stressed animals, can grow more slowly than calmer children. Researchers have known for quite a while that anxious adults often have low levels of growth hormone, and a study in 1997 found that anxious girls, though not anxious boys, were more likely to be short than calm girls.

My guess is that eventually we'll find out anxious boys are smaller, too. Anxious male animals are smaller than calm male animals, and I don't see any reason why human males should be different. I think the German orphanage story probably tells us stress is bad for boys, too. That's the famous case of two orphanages in postwar Germany where one was run by a nice headmistress, while a mean lady who made fun of the children in front of their friends ran the other. She was nice only to the eight children who were her special favorites.

None of the children had enough food, and all of them were smaller than they were supposed to be. Then a natural experiment happened when the government gave the children living with the nice lady extra rations—at the very same moment that the nice lady quit her job and left, and *the mean lady was hired in her place.* The eight teacher's pets moved to the new orphanage with the mean director. Doctors were measuring all the children's growth, and they found that even though the children in the first orphanage were getting extra food, now that they were stressed by a nasty adult they didn't grow as well as the children in the other orphanage. They had more food but grew less. The eight favorites grew better than anyone. Both orphanages had boys as well as girls, so I assume the boys' growth was slowed by stress, too.

With animals there's no ambiguity: stress is horrible for growth, period, which means stress is horrible for profits. So even a feedlot owner who doesn't care about an animal's feelings doesn't like using prods, because a stressed animal means financial loss.

When I got to the feedlot it took me about ten minutes to figure out the problem.

To get to the squeeze chute, first the animals had to walk inside the barn door into a round holding area called a crowd pen. That part of the procedure went off without a hitch. The cattle didn't have any problem stepping inside the pen.

Next they were supposed to walk into a curved single-file alley (it's also called a chute) that led to the squeeze chute. That was where the cattle balked. They just would not walk into the alley. It was the exact same alley feedlots all over the world were using without any trouble, so no one could figure out what the problem was. They couldn't see anything about their setup that was different from any other setup.

But to me it was obvious: the alley was too dark. The cattle were supposed to walk from broad daylight into an unlit indoor alley, and the contrast in illumination was too sharp. They were afraid to walk into pitch-black space.

That might seem a little surprising, since prey animals, like cattle, deer, and horses, usually like the dark. They can hide in the dark and feel safe, or at least safer than they feel during the day. But the problem wasn't the dark, it was the contrast of going from bright sunlight to a dark interior. Animals never like going from bright to dark. They don't like any kind of experience that temporarily blinds them, and that includes looking into a bright light when they're standing in relative darkness. I've found that cattle won't even walk toward a glaring lightbulb. You have to use indirect lighting at the mouth of an alley to make it work.

As soon as I saw the setup I figured that was the problem, and I confirmed my guess when I asked the owner how the cattle behaved at different times of the day, and in different kinds of weather. When he thought about it, he realized that the facility worked fine at night. Things weren't too bad on cloudy days, either. It was the bright, sunny days that were impossible, but no one had noticed the pattern.

I think a number of things are at work when an animal reacts that way. Cattle have excellent night vision and are used to seeing well in the dark, unlike people. So the experience of going temporarily blind in the seconds before their irises expand, which is something people take for granted, probably makes them panic. Also, cows don't live in houses with electricity and drive around in cars at night the way we do, so they don't develop a mental category called "eyes adjusting to an abrupt change in illumination." Last but not least, animals are so intensely sensitive to the visual world that I wouldn't be surprised to find out that sudden huge changes in illumination are physically painful in some way. People don't enjoy the experience of moving from brilliant light to a dark room, either, but for a cow it must be overwhelming.

Maybe when those cattle started to walk out of the sun into the chute they felt like they were going blind for real. They might have been having the same reaction you or I would have if we were driving down the street and suddenly went blind every time we drove through an underpass. If you went blind every time you drove through an underpass you wouldn't drive through underpasses.

I always tell people: whenever you're having a problem with an animal, *try to see what the animal is seeing* and experience what the animal is experiencing. There are lots of things that can upset an animal—smells, changes in routine, exposure to things he hasn't experienced before—and you should consider all of them. Anything in the sensory realm can upset an animal. But don't forget to ask yourself what your dog, cat, horse, or cow may be *seeing* that's bothering him.

At that feedlot, all they needed to do was get more light inside the barn. They could have fixed the problem themselves in five minutes if they'd been able to think about the chute from the animal's point of view. The answer was right in front of them. I really do mean directly in front of them, because the people who built the barn in the first place had installed a big sliding garage door on the front of the barn that the owner had left closed.

When I told him all they needed to do was open the door, it turned out that it hadn't been opened once since the lot was built. They didn't even know if they *could* open it after all this time. But they got a couple of guys to put their shoulders up against the door, and after a few minutes of straining and grunting they got the thing open. That was the end of the problem. The cows all walked into the chute just as nice as could be.

WHAT PEOPLE SEE AND DON'T SEE

That feedlot consultation was the kind of thing that started to give me a reputation for having practically a magical connection to animals. Meanwhile I was always mystified by these situations, because to me the answers seemed so obvious. Why couldn't other people *see* what the matter was?

It took me fifteen years to figure out that other people actually *couldn't* see what the problem was, at least not without a lot of training and practice. They couldn't see it because they weren't visually oriented the way animals and autistic people are.

I always find it kind of funny that normal people are always saying autistic children "live in their own little world." When you work with animals for a while you start to realize you can say the same thing about normal people. There's a great big, beautiful world out there that a lot of normal folks are just barely taking in. It's like dogs hearing a whole register of sound we can't. Autistic people and animals are *seeing* a whole register of the visual world normal people can't, or don't.

I don't just mean this metaphorically, either. Normal people literally don't see a lot of things. There's a famous experiment by a psychologist named Daniel Simons, head of the Visual Cognition Lab at the University of Illinois, called *Gorillas in Our Midst,* that shows you how bad people's visual awareness is. In the experiment they show people a videotape of a basketball game and ask them to count how many passes one team makes. Then, a little while into the tape, while everyone is sitting there counting passes, a woman wearing a gorilla suit walks onto the screen, stops, turns, faces the camera, and beats her fists on her chest.

Fifty percent of all people who watch this video *don't see the gorilla!*

Even when experimenters ask them directly, "Did you notice the gorilla?" they say, "The what?" It's not that they don't *remember* the lady in the gorilla suit. Anyone who's forgotten something he saw will remember it when you give him a prompt. These folks actually didn't see the lady gorilla in the first place. She didn't register.[6]

The experimenters tested out their theory with another video in which an actor suddenly changes into a whole different person, wearing a completely different set of clothes. Seventy percent of normal people don't notice that, either. They also don't notice it in real life. In one study a blond-haired man wearing a yellow shirt handed students a form to fill out, then took the completed form behind a bookcase to file. When he came back out he was a dark-haired man wearing a blue shirt. He wasn't the same guy in disguise; he was a whole different person. It didn't matter. Seventy-five percent of the students had no idea they'd just interacted with two different people.

The scariest study, though, was the one NASA did with commercial airplane pilots. The researchers put them in a flight simulator and asked them to do a bunch of routine landings. But on some of the landing approaches the experimenters added the image of a large commercial airplane parked on the runway, something a pilot would never see in real life (at least, let's hope not). *One quarter of the pilots landed right on top of the airplane.* They never saw it.

I've seen photographs from the study, and what's interesting is that if you're *not* a pilot, the parked plane is obvious. You can't miss it, and you don't have to be autistic to see it, either.[7] I'd bet the ranch that the only people who could possibly miss that plane would have to be commercial pilots. If you're a professional, expecting to see what a professional normally *would* see, there's a 25 percent chance you'll miss a huge commercial aircraft parked crossways blocking the landing strip in a flight simulator.

That's because normal people's perceptual systems are built to see what they're used to seeing. If they're used to seeing gorillas in the middle of basketball games, they see gorillas. If they're not used to seeing gorillas in the middle of basketball games, they don't. They have *inattentional blindness.*

I have no idea how a visual thinker would do on these experiments, but my guess is visual thinkers would see the gorilla a lot more often than verbal thinkers. I'm almost positive there's no prey animal on earth who would miss that gorilla, that's for sure, though I think predators would see the gorilla, too. A *predator,* by the way, is an animal like a dog or a cat who hunts and kills other animals for food; a *prey animal* is the animal the predator hunts. There's also another category of animals you don't hear about as much, which is the *scavenger* animals (like vultures) who do eat meat but don't kill the animals they eat. All animals, including human beings, fall into at least one of these categories, and quite a few—including a lot of primates—belong to more than one. Humans are more predators than prey, but we share qualities with both. In terms of the size of our teeth, we're defenseless, but as soon as we developed tools we became predators.

It's so hard for normal people to see what scares cattle that I finally developed a checklist of mostly visual details for plant managers to look out for. Things like pieces of metal that wiggle, reflections on water, bright spots, contrasts of color, and air hissing or blowing in their faces. I tell the owners, if you have three "bad" details you have to correct *all three*. Then your animal will walk up the chute without any trouble and you can throw away your electric prod.

Visual thinkers of any species, animal or human, are detail-oriented. They see everything and they react to everything. We don't know why this is true, we just know from experience that it is. I've had interior designers tell me, "I see everything." The worst thing that can happen to an interior designer is to work with a sloppy contractor. The designer will see every little flaw in the contractor's work. Tiny mistakes no one else even notices, like grout that's slightly uneven, will jump out at visual people. They go crazy. Visual people feel horrible when little details in their visual environments are wrong, the same way animals do.

I think this is probably the hardest part of an animal's existence for normal people to relate to. Verbal people can't just turn themselves into visual people because they want to, and vice versa.

—

I hope this book will help regular people be a little less verbal and a little more visual. I've spent thirty years as an animal scientist, and I've spent my whole life as an autistic person. I hope what I've learned will help people start over again with animals (and maybe with autistic people, too), and begin to think about them in a different way.

I hope what I've learned will help people *see*.

2. How Animals Perceive the World

The problem with normal people is they're too cerebral. I call it being *abstractified*.

I have to fight against *abstractification* constantly when I'm working with the government and the meatpacking industry. A big part of my job now is trying to make sure all food animals are given a humane slaughter, but even though there's a lot of support for animal welfare it's getting harder to make good reforms instead of easier. It's harder because today government regulatory agencies are all run by people who've been to college, but who in some cases have never even been *inside* a meatpacking plant, let alone worked in one. It's terrible. I keep telling them, "You have got to go out there and visit a plant."

Things were different in the 1960s when I was visiting my aunt's ranch in Arizona. That was my first experience with the United States Department of Agriculture. At that time livestock were being attacked by screwworms all over the West, Southwest, and Mexico. Screwworms are the larvae of a fly that lays its eggs in open wounds. The wounds can be from anything—a cut, a tick bite, or even a newborn's navel. (Screwworms can attack humans, too, and like to lay their eggs inside the nostril.) When the eggs hatch the maggots come out and eat the animal alive. Other maggots eat dead flesh, but screwworm maggots eat live flesh and they are deadly.

Up until the USDA got involved, my aunt had been digging the maggots out of wounds on her horses by hand. She would pick each maggot out with a tweezers, drop it on the ground, and squash and stomp it. Then she'd blob screwworm paste all over the wound to fill it up so no flies could get back in and lay more eggs. The paste looked like black roofing cement. If you didn't do this, the horse would die. A screwworm infestation was a hideous, horrible thing.

The USDA fieldworkers figured out how to get rid of the screwworms by taking advantage of a quirk in their reproductive system. The screwworm's developmental sequence goes from egg to maggot to pupa to fly, and the USDA bred a bunch of screwworms and irradiated the males when they reached the pupa stage, making them sterile. Then they put the pupae in little paper boxes, like a Chinese takeout box, and dropped the boxes out of airplanes. The flies would come out of the boxes and mate with lots of females, and the females they'd mated laid eggs that didn't hatch.

The program was a huge success. It started in 1959, the United States working with Mexico, and the last case of screwworm infestation was recorded in Texas in 1982. Today there are no screwworms anywhere in the United States or Mexico. I remember those years well. You'd find the little boxes all over the ranch, seven or eight of them each summer. The box would say "USDA" and there would be a little story printed on the side explaining what it was and that it wasn't going to hurt you.

This was the original biotechnology and it worked. The government saved thousands and thousands of animals, maybe millions. They just did it; they didn't get everyone's permission.

Today the government could never get a program like that off the ground. Some environmental activist would say, "We have to protect these flies," and you'd have people who'd never seen a screwworm in their lives advocating to save them from extinction. The whole thing would be about ideology, not reality. The USDA would be required to file environmental impact statements and the environmental impact statements would be challenged in court, and it would never get done.

Even worse, the government might not even get to the point of having advocates block their efforts. To put this type of project together you need a really good field staff that is *in charge of things*. But today the abstract thinkers are in charge, and abstract thinkers get locked into abstract debates and arguments that aren't based in reality. I think this is one of the reasons there is so much partisan fighting inside government. In my experience, people become more radical when they're thinking abstractly. They bog down in permanent bickering where they've lost touch with what's actually happening in the real world. The only way anything can get done is when there's an emergency. Then all of a sudden everyone has to move.

—

So the 1960s and the 1970s were the golden age; that was a time when people who were in charge of regulation, or who were running the plants, had actually done things with their hands.

One thing I've noticed about animal welfare regulators who have never worked in the industry is that they always go for some kind of zero-tolerance approach. If the plant violates one or two agency rules, it has to be shut down.

If you don't know anything about the meatpacking business, that sounds like a good idea. Make sure no animal *ever* gets hurt, under any circumstances.

But in real life that's never the way it works out. In real life what happens is that a plant makes one or two mistakes, so the agency shuts it down. Well, shutting down a plant creates a huge uproar, because you've closed a whole big huge company that employs a lot of people. Management immediately protests the decision, and lots of pressure gets put on the inspector who reported the violations to clean up his report so the plant can go back to work.

And that's what happens. The plant goes back to work and doesn't get inspected so closely anymore. The violations keep on piling up.

It doesn't have to be that way. I constantly argue that what we really need to do to protect animals is *set high standards*. People can live up to high standards, but they can't live up to perfection. When you give a plant a good standard—like 95 percent of all cattle have to be stunned (killed) correctly on the first shot every single day—they always do better than they do under zero-tolerance regulation. A lot of times they beat the standard, too.

But regulators today are too abstract in their thinking to see that. They're focused on their *thoughts* about the animals, not on the real animals in the real plants, so more animals end up suffering. It's not right.

HOW PEOPLE SEE THE WORLD

Unfortunately, when it comes to dealing with animals, all normal human beings are too abstractified, even the people who are hands-on. That's because people aren't just abstract in their thinking, they're abstract in their seeing and hearing. *Normal human beings are abstractified in their sensory perceptions as well as their thoughts.*

That's why the workers at the facility where the cattle wouldn't go inside a dark building couldn't figure out what the problem was. They weren't seeing the setup as it actually existed; they were seeing the *abstract, generalized concept of the setup they had inside their heads.* In their minds their facility was identical to every other facility in the industry, and on paper it *was* identical. But in real life it was different, and they couldn't see it. I'm not just talking about management. The guys in the yard, who were there working with the animals, trying to get them to walk inside the building, couldn't see it, either.

That's the big difference between animals and people, and also between autistic people and nonautistic people. Animals and autistic people don't see their *ideas* of things; they see the actual things themselves. We see the details that make up the world, while normal people blur all those details together into their general concept of the world.

A huge amount of my consulting business is getting paid to see all the stuff normal people can't see. I do this constantly. Not too long ago I got a call to go out to a meatpacking plant where the animals were getting big fat bruises on their loins. The loin is the area in between a cow's rib cage and its rear leg. It's the most expensive part of the animal, because that's where the steak is located. So nobody wants their cattle getting bruised loins. A bruise means bleeding inside the muscle, and the bloody area has to be cut out in the butchering process, which means less meat to sell. Delaying slaughter until the bruise clears up doesn't help, either, because a healed bruise leaves behind tough meat and gristle. Gristle is scar tissue. Just about any injury, no matter how tiny, can produce gristle, including the needle used in a cow's vaccinations. (To prevent scarring from vaccination, you have to give the shot just under the skin.

The beef industry is working hard trying to get feedlot employees and ranchers to give shots correctly.)

So here was this plant with all its beautiful, well-tended cattle walking around with big bruises on their sides, and nobody could figure out how they were getting them. One minute a cow would be fine; the next minute the same cow would have a great big shiner on her side.

They brought me out, and I walked into the chute to take a look around. That's the first thing I always do, because you can't solve an animal mystery unless you put yourself in their place—*literally* in their place. You have to go where the animal goes, and do what the animal does.

The chute turned out to be the problem. There was a sharp three-inch piece of metal sticking out from the side, and the cattle were hitting it. That little shard of metal was obvious to me, but not one person at the plant had spotted it—and all of them were looking. I think they probably would have seen it pretty quickly if any of the cattle had bellowed when they hit it, but the cattle didn't yelp. The animals were hitting hard enough to bruise themselves, but not hard enough for it to really hurt.

WHAT DO ANIMALS SEE?

When an animal or an autistic person is seeing the real world instead of his idea of the world that means he's seeing *detail*. This is the single most important thing to know about the way animals perceive the world: animals see details people don't see. They are totally detail-oriented. That's the key.

It took me almost thirty years to figure this out. During all that time I kept a growing list of small details that could spook an animal without realizing that "seeing in details" was a core difference between animals and people. The first small detail I saw spook a cow was shadows on the ground. Cattle will balk at the sight of a shadow. Then the workers get out the electric prods, because they have no idea what's scaring the cattle, so they can't fix it. I first saw cattle get spooked by a shadow thirty years ago, and I've been seeing it ever since.

The next detail I noticed was that cattle were afraid to enter dark places. That got me on the track of thinking that differences in contrast were important for animal behavior, which is true, but it didn't tell me that detail per se was the issue.

I finally realized that animals perceive way more details than people do when McDonald's hired me in 1999 to help them implement the animal welfare audit I'd originally created three years earlier for the USDA. They had a list of fifty meatpacking plants they purchased beef from, and they had announced that all fifty plants had to pass my audit or get thrown off the list.

McDonald's was already auditing their suppliers for food safety, so they asked me to train their auditors to monitor animal welfare, too. It was easy to train the auditors, but it wasn't easy for all the plants to get in compliance, even though they wanted to. Good intentions weren't enough. We had to help plants figure out what they were doing wrong.

One of the criteria the plants had to meet to pass my audit was that employees couldn't use the electric prod on more than 25 percent of the animals. Any plant that *couldn't* get its prod usage down to 25 percent had to analyze what the problem was and correct it. But sometimes no one at the plant could see why their animals were balking.

Always, when I would go out to the plant to analyze the situation, I would find two things.

First, the problem was always a *small detail,* usually a detail the humans hadn't even noticed. The entrance to the chute might be too dark, or there might be a bright reflection on a metal bar that was causing the animals to balk.

Second, to get their prod scores down a plant had to correct *all* the details that were scaring the cattle. They couldn't just correct some of the details or most of the details. They had to correct all of the details.

There was this one hog plant on the list that had four things they had to fix. Three involved lighting and the fourth was that they needed to put up some metal sheeting to prevent the pigs from seeing people moving around up ahead. This is something most people don't realize: cattle and hogs raised for food are domestic animals, but they aren't naturally tame unless they've been socialized to humans as babies. So they get jittery when they're walking through a chute or alley and see people moving up ahead of them. All domestic animals, including cats and dogs, have to be socialized to people. The plant had to make all four corrections to get their prod scores down. They couldn't just fix three and let it go at that.

That turned out to be true at all the plants. No plant had zillions of bad details; about the most any plant had was six. But if they had four bad details they had to correct all four. For the animals every detail was equally bad and equally important. That's what made me realize that details are the key, and that's when I started preaching the importance of detail in all my talks and all my articles and books.

Only highly visual people react to details the way animals do. I knew one interior designer who was supervising a renovation of her own bathroom and the contractor cracked one of the marble tiles. She couldn't stand it. Every time she went in the bathroom she saw that crack. It jumped out at her and she'd get upset all over again. She knew she was different, but that's what made her good at her job. She saw the visual details most people didn't.

Nancy Minshew, a research neurologist at the University of Pittsburgh who specializes in autism, was coming out with her new work on autistic people's cognitive processing around the same time, and she confirmed my new insight into animals and detail. Her brain scans showed that autistic people are much more focused on details than on whole objects. Since I'd noticed so many similarities between animals and autistic people in my career, the fact that Nancy Minshew was finding a connection between autism and an orientation to detail gave me another reason to think I was right about animals.[1]

TINY DETAILS THAT SCARE FARM ANIMALS

Here's the checklist I give plant owners when their cattle or hogs are refusing to walk through an alley or a chute:

1. SPARKLING REFLECTIONS ON PUDDLES

I figured this out at a plant where the pigs were constantly backing up in the alley, so the employees were using electric prods to keep them moving forward. The plant was failing its animal welfare audit, because workers were supposed to be using the prods on no more than 25 percent of the pigs, and they were using them on every single animal. Normally a pig has no problem walking through a chute, but in this plant every single pig was stopping and backing up.

I got down on my hands and knees and went through the chute the same way the pigs did. The managers probably thought I looked crazy, but that's the only way you can do it. You have to get to the same level as the animals, and look at things from the same angle of vision.

Sure enough, as soon as I got down on all fours I could see that there were lots of tiny, bright reflections glancing off the wet floor. Plant floors are always wet, because they're always being hosed down to keep them clean. Nobody could have seen those reflections even if they did know what to look for, because the humans' eyes weren't on the same level as the pigs'.

Once we knew what the problem was I got back down on my hands and knees again, and while I was pretending I was a pig the employees moved the big hanging lights overhead with a stick until each little reflection was gone. And that was that. Once the reflections were gone the pigs walked right up the chute, and the plant passed its audit.

2. REFLECTIONS ON SMOOTH METAL

I first saw this with cattle walking up a single-file chute that was made of shiny stainless steel. Every time the sides jiggled the shiny reflections from the lights would vibrate and oscillate, and the cattle would stop. In that plant all we had to do was move the lights, but in another plant with the same problem, we had to bolt the sides down so they couldn't move at all.

A still reflection is always less of a problem for an animal than a moving one, although any bright reflecting surface can scare an animal. A lot of times we have to move the lights *and* bolt down the metal sides. A number of things can cause reflections to move: machine vibrations, or cattle banging up against the metal, or water running off a ramp into the water that's already on the floor, making the reflections on the surface jump and move like a sparkling brook.

3. Chains That Jiggle

I learned about jiggling chains in a big beef plant in Colorado that had a chain hanging down at the entrance of the chute. The chain was part of a gate latch, and it wasn't very long; maybe only one foot, and swinging back and forth three inches each way. But that was enough. The cattle would come around a curve, take one look at that chain, then stop and stare at it with their heads swinging back and forth in rhythm with the chain. You'd think that would be obvious to the employees, but it wasn't. The humans just didn't see it, even though the cows' heads were going back and forth in rhythm to the swinging of the chain. I'm not sure the employees even noticed that the cows' heads were moving; forget the chain. The employees were just using more force, zapping them with cattle prods, screaming and yelling and so on, to try to get the cattle moving.

4. Metal Clanging or Banging

This one's universal. You see it everywhere in feed yards and plants—metal gates, sliding doors, squeeze chutes—everywhere. People in the industry call it *clatter*, and clatter is something you always have with metal equipment. I recommend plastic tracks for sliding doors, so you don't have metal sliding against metal, and now a company named Silencer makes an extra-quiet squeeze chute that's good, too.

5. High-Pitched Noise

Examples: backup alarms on trucks and high-pitched motor whining.

I remember my first experience with this at a big beef plant in Nebraska where they'd just put in one of my cattle-handling systems. They used a hydraulic system that gave off a high-pitched whining noise, and the noise would get the cattle all agitated so my system didn't work. We changed the plumbing to eliminate the noise and the cattle became a lot calmer.

6. Air Hissing

Another one you see everywhere. The problem with high-pitched sounds like hissing air and hydraulic squeals is that they're too close to distress calls, which are almost always high-pitched. High-pitched sounds are one of the few things humans will usually notice, especially if they're intermittent, because we inherited a built-in alarm system from our animal ancestors that's still working. That's why humans choose high-pitched intermittent sounds when they want to make sure they get people's attention. Police cars, ambulances, garbage truck backup beeps—it's almost always a high-pitched inter-mittent sound. The people who design these systems instinctively go for the kind of sound animals use to signal danger.

7. Air Drafts Blowing on Approaching Animals

I don't know why cattle don't like this; I just know they don't. Whenever cattle are out in a big storm, they'll turn their bottoms to the wind. I also hear stories about dogs hating to have air blown into their faces or their ears. This seems to be something kids like to do to dogs, so I've heard quite a few of these stories.

8. Clothing Hung on Fence

I say "clothing" because the problem almost always is clothing, but anything hanging on a fence can scare animals. Usually what happens is that people get hot, take off their jackets and shirts, and hang them on the fence. Sometimes people will drape towels or rags on the fence, which is just as bad. Once I went to a ranch that had a wiggling plas-tic jug wired to the fence and that was causing problems.

The worst is when you have yellow clothing hanging on fences. I first saw this happen at a plant in Colorado. It's the same problem as the bright yellow ladder against the gray wall I mentioned a while back. No cow will walk toward a sudden patch of bright yellow color.

9. Piece of Plastic That Is Moving

Anything moving is a problem for animals, but usually I find the problem will be a piece of plastic. That's because people in the industry put plastic all over everything. They'll tape it over a window to keep the cold air out, or wrap it around a pipe because the pipe is dripping, and it always vibrates and jiggles. Plastic just has a way of getting stuck all over the place, especially now, with the new food safety rules. Employees pull plastic off big rolls and make raincoats out of it, or aprons and leg guards; the plants let the employees make anything they want out of the stuff. Then it ends up getting caught on something where it jiggles and scares the animals. Paper towels will also scare pigs and cattle if the wind is blowing it. I had a paper towel problem at five or six different places.

10. SLOW FAN BLADE MOVEMENT

I've seen this in several different places. Animals don't have a problem with an electric fan that's *turned on* the way autistic children do. A lot of autistic children are riveted by the motion of the blades, or by just about anything that's spinning fast. I don't know why this happens, but I think they may be seeing the flicker of the fan blades even at very high speeds. I've met a number of dyslexic people who can see the flicker, so I assume many autistic people see it, too. Dyslexics who can see the blade flicker say it's horribly distracting and fatiguing.

The motion is part of the attraction, too. I don't get hooked on fans myself, but I do get stuck on those geometric screen savers a lot of computers have. I can't stop looking at them, literally, so if I'm in an office where there's a geometric screen saver either I have to sit with my back to the screen, or ask the owner to turn it off.

With fans, what drives an animal crazy is when the fan is turned off, but the blades are rotating slowly in the breeze. You have to put up big pieces of plywood or metal so the animals can't see the fan. Otherwise, forget it. They're going to balk. I went to one ranch where they had a windmill that was messing up the animals. On windy days the animals wouldn't move.

11. SEEING PEOPLE MOVING UP AHEAD

Another case for plywood. I mentioned this one earlier. Cattle are eighteen months old when they're slaughtered, and pigs are only five months old, so it doesn't pay to *train them to lead*. They're not like horses who've been trained to accept a halter and a lead rope and walk calmly alongside a human being.

12. SMALL OBJECT ON THE FLOOR

Example: a white Styrofoam coffee cup on a muddy brown floor.

I had a bad experience with this one time when I was up on a cat-walk above a cattle chute. An employee at the plant had been storing his white plastic water bottle on the catwalk, and I accidentally kicked it off. The minute it hit the ground, I said a bad word. It landed right at the entrance to the chute, where I knew it was going to cause a problem, and it did. That little plastic water bottle lying harmlessly on the ground was as big a barrier for those 1,200-pound cows as if I'd dropped a big pile of boulders there.

We had to shut the whole line down, because no animal would walk over it, and it was too dangerous for anyone to go in there and try to pick it up. A crowd pen is a small space, and there were fifteen big animals in it, none of them trained to lead; a human going inside the pen could have been crushed. So the employees had to stand outside and run at the cattle and chase them until finally one of the cows stepped on the bottle and crushed it into the manure so that it turned brown, not white. Then the cattle were fine. They all stepped over it and went on into the alley. That part of the line was shut down for fifteen minutes, and the plant as a whole lost five minutes. At $200 a minute that was a $1,000 delay.

13. CHANGES IN FLOORING AND TEXTURE
Example: cattle or pigs moving from a metal floor to a concrete floor or vice versa.

The problem is contrast.

14. DRAIN GRATE ON THE FLOOR
Same problem again: contrast. The drain grate looks too different from the floor.

15. SUDDEN CHANGES IN THE COLOR OF EQUIPMENT
High-contrast color changes are the worst. You can't have the gates painted one color and the pens painted another. I've also seen problems with gray-painted alleys leading up to shiny metal equipment.

16. CHUTE ENTRANCE TOO DARK
Another contrast issue—going from light to dark.

17. BRIGHT LIGHT SUCH AS BLINDING SUN
If you have the sun coming up over the top of a building just as the cattle are approaching there is nothing you can do. It is a hell of a problem

and there isn't any way to fix it except maybe extend the roof out over the yards. Otherwise you just have to suffer through it.

18. ONE-WAY OR ANTI-BACKUP GATES

These are two different terms for the same thing. *Anti-backup gates* don't look like the normal gates the cattle are used to seeing on a ranch. Anti-backup gates hang down from overhead instead of being attached on one side, and basically look like a cow- or pig-sized dog door in a house. Plants install one-way gates in single-file alleys to keep the cattle from backing up into the long line of animals behind them. The pig or cow pushes through the gate—the same way a dog pushes through a dog door—and the gate falls down behind each pig or cow after it walks through. It's not flexible like a dog door, so you can't push it backward, only forward.

The animals hate having to push through the gate. That's the problem, the going-through. The anti-backup gates bother the animals so much I don't like to use them. I work with the cattle gently enough that they're all happy to keep walking forward, and I can just tie the doors up out of the way, where the cattle don't see them and don't have to deal with them.

—

You could make up the same kind of list for any animal, although it would be different for each one. Bats have sonar and dogs don't, so the list of common distractions for bats is going to have some sonar distracters on it, while the dog's distracter list won't. But any list of common distractions for an animal would be highly, highly detailed, exactly like this one.

THE DIFFERENCE BETWEEN ANIMAL VISION AND HUMAN VISION

Although I created this list for cattle and hogs, you can use this list to predict trouble spots for any other animal if you think about what these eighteen distracters have in common.

First of all, fourteen out of the eighteen distracters are visual, and I wouldn't be surprised to find a ratio like that for most animals. But to predict *what kind* of visual object will distract or frighten an animal, you have to know more about what animal vision is like.

It's pretty different from ours. For instance, you always hear that dogs "don't see well," which is true as far as it goes. Dogs don't have very good visual acuity, which is the ability to see the tiny details of what you're looking at clearly and crisply. People with 20/20 vision have excellent visual acuity, and a lot of animals don't. That means that most animals aren't going to be frightened by tiny objects, simply because they can't see them well.

A typical dog has a visual acuity of 20/75, which means that a dog has to stand twenty feet away to clearly see an object a person with normal vision sees well standing seventy-five feet away. The dog has to get much closer to the object than we do. This isn't due to nearsightedness but to the fact that dogs have fewer cones in their retinas than people do. Everyone probably remembers from biology class that cones handle color and daytime vision, and rods handle nighttime vision. Basically dogs have traded good visual acuity for good nighttime vision. A dog doesn't see *any* objects as sharply as a person does, including an object that's right under his nose. That's why it's so hard for dogs to see a piece of kibble you've dropped on the floor for them to eat. If they didn't watch it fall, most dogs can't see it lying on a mottle-colored tile floor (though some can).

There's also a lot of variation in visual acuity among the different breeds of dogs, as well as among individuals of a breed. One study found that 53 percent of German shepherds and 64 percent of Rottweilers were nearsighted. You might wonder whether being nearsighted matters to a dog since everything it sees is fuzzy to start out with, but tests show that it does. A nearsighted dog has much worse visual acuity than a normal-sighted dog. Interestingly, although German shepherds tend to be nearsighted, only 15 percent of the Shepherds in a demanding program for guide dogs were myopic.[2] Probably the nearsighted dogs were flunking out of the program without the trainers' knowing why.

Another huge difference between animals and people is that most animals have panoramic vision. The eyes of prey animals like horses, sheep, and cows are set so far apart that they can literally see behind their heads. That's why some hansom cab horses wear blinkers; they can see everything going on behind them, and they get distracted. Most racehorses *don't* wear blinkers for the same reason: their trainers want them to know exactly where the horses behind them are, and how fast they're moving.

Prey animals don't have perfect 360-degree vision, although they come close. There's one small blind spot directly behind a cow or horse that you have to be careful not to sneak up to. The animal can't tell what you are, and he might get scared and lash out and kick you. Prey animals also have a small blind spot directly in front of their heads because their eyes are set so far to the sides.

Even though their eyes are so far apart, prey animals *do* have depth perception, though it seems to be different from ours. We use *binocular vision*, which means each eye is seeing the same thing from a slightly different angle. When our brains combine the angles, we get our sense of depth.

Prey animals' eyes are so far apart that a lot of researchers have assumed their left eye was seeing something completely different from the right eye, so they couldn't have binocular vision. But they've tested this in sheep, and sheep do have at least *some* binocular vision. We know this because sheep can see the cliff in *visual cliff* experiments. In the original visual cliff studies the experimenters put a baby on top of a table covered in a sheet of glass thick enough to crawl on. Directly underneath the glass there was a checkered surface that, midway across the table, suddenly dropped off way below the glass surface. It was a *visual* cliff, not a real one, so the baby couldn't actually fall over the edge if he crawled out over the dropoff. Very young babies will refuse to crawl over the cliff even if their mothers stand on the opposite side of the table and call them. They can see the cliff, and they instinctively know it's dangerous. It turns out that sheep won't walk over the cliff, either, which means they have to be seeing the difference in depth. (On the other hand, sheep don't appear to have depth perception while they're moving, only when they stand still.)

You've probably seen bulls in bullfights lower their heads before they charge the matador. Border collies do the exact same thing when they're herding sheep. They lower their heads below their shoulders and stare at the sheep. They do this because their retinas are different from ours. The human retina has a fovea, which is a round spot in the back of the eye where you get your best vision. Domestic animals and fast animals who live on the open plains like antelopes and gazelles have a *visual streak* instead of a fovea. The visual streak is a straight line across the back of the retina. When you see an animal lower its head to look at something, it's probably getting the image lined up on its visual streak. Most experts think the streak helps animals scan the horizon.

Researchers have also found that of the meat-eating animals that have been tested so far, the two fastest animals—the cheetah and the greyhound—also have the most highly developed visual streaks. Their visual streaks are dense with photoreceptors, giving them extra-acute vision. To test visual acuity you can use a bar code design. The more acute your vision, the tinier a bar code you can look at, from a greater distance, and still see the stripes as separate rather than as a gray square. Animals with super-acute vision can also see separate grains of sand on the beach.

SEEING COLOR AND CONTRAST

A third area where animals and people diverge is in the ability to see color and contrast. At least ten of the eighteen distracters are high-contrast images, like a shiny reflection on metal, or a sparkling reflection in a puddle. Several of the other visual distracters, such as a white Styrofoam or plastic coffee cup on the floor or a piece of clothing hanging over a fence, involve contrast, too. I have some photographs of high-contrast distracters on my Web site. One is a picture of a white coffee cup on a brown floor; another is a pair of bright yellow boots against a gray floor and railing.

Sharp contrasts are also a problem when you're trying to move an animal toward an area that's either too dark or too light. We already talked about the cattle that wouldn't go into the squeeze chute building because it was too dark, but cattle will also refuse to walk directly into an area that is too bright. Strong changes in light are so distracting to cattle that you can't have direct sources of lighting, like an unshaded lantern or lightbulb, at the mouth of an alley. They won't walk toward it. You want overhead lighting with no shadows, like the light outdoors on a bright but cloudy day. Sometimes you can get that effect with skylights made out of white translucent plastic.

Slowly rotating fan blades are also a high-contrast stimulus, because animals see contrast differently from the way we do. If the fan is turned on and is rotating so fast you can't see the blades, there's no problem. But when a fan blade is turning slowly it creates a flicker, and that flicker is a much higher contrast image for an animal than it is for us.

Animals see more intense contrasts of light and dark because their night vision is so much better than ours. Good night vision involves excellent vision for contrasts and relatively poor color vision. I first learned about animals' incredible contrast vision back when I was taking black-and-white pictures of the cattle chutes. There'd be a shadow on the ground that even I wouldn't see until I got the pictures developed. The reason I could see it only in my photographs is that contrast is much sharper when you take away color. Shadows are so much clearer in black and white that during World War II the Allies recruited people who were completely color-blind—not just red-green color-blind, but people who didn't see any color at all—to interpret reconnaissance and spy photos. They could spot things like netting draped over a tank to camouflage it that were invisible to people whose color vision was normal.

Animals seem to see sharp contrast on the floor as a false visual cliff; they act as if they think the dark spots are deeper than the lighter spots. That's why *cattle guards* work on roads. A cattle guard is a pit dug across a road, covered with metal bars. A car can drive over it and a cow *could* walk over it if it tried, but it won't because it sees the two-foot drop-off between the bars.

To a cow the contrast is so sharp the drop-off probably looks like a bottomless pit. In *An Anthropologist on Mars* Oliver Sacks has an essay about an artist who lost his color vision in a car crash. After that it was hard for him to drive, because tree shadows on the road looked like pits his car could fall into. Without color vision, he saw contrasts between light and dark as contrasts in depth.[3] Since cows have much poorer color vision than normal people do and mainly see colors in the yellow-green range, they may see light-dark contrasts as contrasts in depth in an analogous fashion to Dr. Sacks's color-blind artist.

Whatever the reason, cows act like Dr. Sacks's color-blind artist. Cattle guards are expensive to build, so a lot of times the Department of Transportation just uses a standard line-painting machine (that's the machine they use to paint the center line on highways) to paint batches of bright white lines across the highway going in the same direction as a crosswalk. It's a poor man's cattle guard.

When the cattle aren't highly motivated to cross the road, a grouping of twenty white lines painted six inches apart will make them stay put, because the contrast scares them. If the cattle are highly motivated, it's a different story. If you've got mama on one side and baby on the other, painted lines won't work. Or if cattle are starving, they'll cross the lines to get to better grazing on the other side of the road. But under normal circumstances, painted lines work just fine.

You need to know something about animals' color vision to predict what visual stimuli they'll experience as high-contrast. The breakdown is pretty simple: birds can see four different basic colors (ultraviolet, blue, green, and red), people and some primates see three (blue, green, and red), and most of the rest of the mammals see just two (blue and green). With *dichromatic,* or two-color, *vision* the colors animals see best are a yellowish green (the color of a safety vest) and bluish purple (which is close to the purple of a purple iris). That means that yellow is the high-contrast color for almost all animals. Anything yellow will really pop out at them, so you have to be careful about yellow raincoats, boots, and machinery.[4]

THE REAL PROBLEM IS NOVELTY

Any sharp contrast between light and dark will draw the attention of a dichromatic animal, either distracting or scaring him. If he's a big animal who you're trying to move from Point A to Point B, a sharp contrast in light and dark will stop him in his tracks.

However, not all high contrast will scare an animal, only high-contrast visual stimuli *that are novel and unexpected*. If dairy cattle are used to seeing bright yellow raincoats slung over gates every day when they enter the milking parlor there'd be no problem. It's the animal who's seeing a bright yellow raincoat slung over a gate for the first time at a slaughter plant or feedlot who's going to balk. Novelty is the key.

The anti-backup gates used in many cattle alleys have the same problem: the cattle have never seen them before, so they don't want to go through them. Novelty is a huge problem for all animals, all autistic people, all children—and just about all normal grown-ups, too, though normal adults can handle novelty better than animals, autistic people, or kids. Fear of the unknown is universal. If you've never seen something before, you can't make a judgment about it; you don't know if it's good or bad, dangerous or safe. And your brain always wants to make that judgment; that's how the brain works. Researchers have found that even nonsense syllables spark positive and negative emotions; to your brain, there's no such thing as neutral. So if you can't tell what something is, you get anxious trying to decide whether it's good or bad.

Any novel object or image in a cow's visual field will get her worried, and if you happen to be trying to move her in the direction of the novel object or image, forget it.

It's different when you don't try to force things. On its own, an animal will always investigate a novel stimulus, even though new things are scary. I learned that back when I was writing stories and taking photographs for *Arizona Farmer Ranchman Magazine*. I noticed that if you just left a pile of camera equipment alone in the middle of the field, all the cows would come up to it and investigate. But if you walked toward them *carrying* the same equipment, they'd take off. Motion was a problem, so if I just stood there holding the equipment, the cows would come to me.

I also noticed that if I got down low to the ground I was a lot less scary to them. At first I was just trying to get the cow's head framed against the sky, without any grass showing in the frame, so I'd crouch down to get the shot I wanted. But then I noticed that when I crouched down, I could get close-ups of the cattle because they wouldn't run away. Those photos were beautiful—big Black Angus heads silhouetted against the blue sky.

Finally one day I decided to just lie down flat on my back and see what happened. They all came up to me and sniffed and licked and sniffed and licked. These were feedlot cattle who weren't tame.

When a cow comes up to explore you, it's always the same. They'll stretch out their heads toward you and sniff you; that's always first. Then the tongue will reach out and just barely touch you, and as they get less afraid they'll start licking you. They'll lick your hair and chew on it, and they like to lick and chew your boots, too. I usually don't let them lick me on my face because cattle have extremely rough tongues and I could get a scratched cornea, although I sometimes just close my eyes and let them go ahead. I don't mind if the tongue goes down my neck. That's okay. And I let them lick my hands. I think they probably like the taste of the salt on your skin.

Sometimes I'll kiss them on the nose.

I wasn't the only person to figure out that it's perfectly safe to lie down in the middle of a bunch of thousand-pound untamed animals. In the 1970s there were a lot of Mexicans coming over the border to work in the feedlots, and when the Border Patrol came around the Mexicans would hide inside the corrals, with the cattle. Five guys would lie down on the ground with a hundred head of Brahman steers surrounding them. Brahmans are the big huge cattle with the hump on their back. They're nice animals, as long as you treat them well, but they're scary-looking to anybody who doesn't know cattle, so the Border Patrol guys wouldn't dare go in those pens.

But it never came to that, because the Border Patrol people never saw any of the illegal workers lying underneath all those cattle. The Mexicans had to lie perfectly still, because if they moved the cattle would run and give them away. And, of course, that would have been really dangerous for the five guys lying on the ground. You don't want a thousand-pound Brahman steer and his ninety-nine friends stepping on you by accident when they're trying to get away. It sounds dangerous, but I don't remember a single person ever getting hurt.

The reason cattle will approach something novel under their own steam is that they're curious. All animals are curious; it's built into their wiring. They have to be, because if they weren't they'd have a lot harder time finding what they need and avoiding what they don't need. Curiosity is the other side of caution. An animal has to have some drive to explore his environment in order to find food, water, mates, and shelter. People say curiosity killed the cat, and that's probably true; curiosity can get an animal into a lot of trouble. But an animal or a person can be too cautious, too. If you're too cautious to explore things, you miss out on things you need.

Being too cautious might make you miss signs of danger, too. Animals and people need to avoid trouble before it happens, and one way to do that is to pick up on signs of danger and act on them *now*, instead of waiting until you're face-to-face with a hungry wolf and then trying to get away. Curiosity drives an animal to explore its environment for signs of danger.

So it makes sense that a cow would voluntarily explore a yellow raincoat hanging on a fence but dig in his heels if you try to *force* him to walk past one. Since anything new could be dangerous, an animal wants a clear escape route before he's going to poke his nose into something he's never seen before. When he's being forced through a one-way alley, there's no escape. So he refuses to move.

—

You can use the exact same checklist with horses, too, partly because they're prey animals like cattle and partly because their lives and environments are pretty similar. Since I spend most of my time with cattle I don't have a good checklist of details that scare dogs or cats, but I *can* tell you that the same principle applies even though they're predators and don't have as many natural enemies to worry about. All animals, predator or prey, have a built-in sense of caution that is triggered by new things.

With dogs, it's a little hard to predict which new things might scare them, since dogs live with people and get exposed to so many new things all the time. A dog who's not naturally timid can *seem* like he doesn't mind high-contrast novel stimuli the way a cow does.

But I don't think that's true. One of the good times to see the effects of novel visual stimuli on a dog is Halloween. My experience is that *dogs do not like Halloween costumes!* A friend of mine was sitting in her upstairs office one day, getting some work done, with the family Lab lying next to her, when her son walked up the stairs wearing his Scream costume. You probably know the one I mean: the costume is dark black, and the mask is bright white with a big red tongue hanging out of its mouth. You can't get much higher contrast than that, unless you made the tongue yellow. The Lab jumped to her feet and started barking her head off.

My friend was totally surprised, because she had recognized her son from his footsteps, which sounded the same way they always did. He wasn't wearing a costume on his *feet*. But the minute the dog saw the mask she went nuts.

This is another example of the cardinal rule of my checklist: just *one* of these distracters, out of eighteen, will throw an animal off. To the Lab, it didn't matter that my friend's son still sounded and smelled the same. He didn't look the same, so he wasn't the same, and that was that. Apparently animals use an additive system rather than an averaging system when they're figuring out what something is and whether they should be afraid of it.

That same Lab also went crazy when the neighbors put a Halloween scarecrow up in the front yard. My friend was taking her dog for a walk when they spotted the scarecrow, and the Lab started barking ferociously at the thing. Her hackles were up, too. That same house managed to throw my friend's other dog into a panic with a piece of lawn sculpture they put in the backyard. The sculpture was a foot-high all-black iron frog, and when the other dog caught sight of it he had the same reaction his pack mate did to the scarecrow. He went nuts. Frantic barking, hackles up, straining at the leash.

Dog and cat owners won't have any problem recognizing the next category of common distracters: things that are moving. For any animal you can name, sudden movement is *riveting*, especially sudden rapid movement. Rapid movement stimulates the nervous system. It makes prey animals run away, and it makes predator animals give chase. It always grabs your attention. That's why used car lots put flags or twirly plastic thingies up all around their lots. You can't *not* look at a bunch of brightly colored, rapidly moving objects. Jiggling parts on feedlot equipment trigger a cow's inborn impulse to flee, and all of a sudden you've got a whole herd of cows turning into the feedlot version of a forty-car pileup. It's a disaster.

SOUND

Last but not least, you have your sound distracters. Any novel, high-pitched sounds will cause cattle to balk, because they activate the part of an animal's brain that responds to distress calls. An *intermittent* high-pitched sound is that much worse. Intermittent sounds will drive anyone crazy; they're much more upsetting than a constant, loud din, whether it's high-pitched or not. You can't relax, because you're waiting for the next sound. And you can't turn this response off, either, because intermittent sounds activate your *orienting response*. People aren't so aware of this response in themselves, but if you live around animals you know it well. Anytime an animal of any species hears a sudden sound, something they weren't expecting, they stop what they're doing and orient to the source of the sound.

When I worked with pigs at the University of Illinois I saw the orienting response every time a small plane would fly over the farm. The pigs couldn't see the plane from inside the barn, but the minute that plane could be heard approaching the farm all activity in the barn would stop dead, and every animal stood perfectly still. After about two seconds of focused listening the pigs went back to their normal hubbub of activity. You can see the same thing at a horse stable when a garbage truck backs up to the dumpster. As soon as the backup warning starts beeping every horse will stick its head out of the stall at the exact same moment and stand at the alert. They look like they're saluting the truck.

I think the orienting response is the beginning of consciousness, because the animal has to make a conscious decision about what to do about that sound. If he's a prey animal, should he run? If he's a predator, does he need to chase something? A predator might need to flee, too, of course, so a predator actually has two decisions to make.

Intermittent sounds keep hitting that orienting response. That's why it's impossible to get to sleep when you're hearing an intermittent sound like a beeping elevator in a hotel or an intermittently beeping clothes dryer. A friend of mine with a nine-year-old autistic boy told me a story about her son, who had gotten into opening and closing doors repetitively. She was exhausted one day, mostly because her son didn't sleep well at night, and she needed to take a nap, but when she lay down her son started opening and closing the sliding pocket door to the laundry room next to her bedroom. He would wait a few seconds in between each new door closing, just long enough for her to start to drift off to sleep again, and then suddenly she'd hear a rumble-rumble-*thump* and the door would hit the doorjamb again. Even though the sound was muffled, she said she was frantic after about ten minutes of this. It's the Chinese water torture principle. If you had water pouring on your head continuously you wouldn't like it, but you could learn to ignore it. Having drops of water dripping on your head intermittently is literally torture.

BEING OBLIVIOUS

The funny thing about the checklist is that probably the *only* thing on it that would bother a herd of humans you were trying to move through a feed yard chute is the intermittent sounds. Humans wouldn't bat an eye at anything else on the checklist—jiggling chains, sparkling puddles, shiny spots on metal, little pieces of moving plastic, slowly rotating fan blades, even a continuous high-pitched sound—nothing on this checklist would be any problem for human beings at all.

They wouldn't be a problem for humans, because humans wouldn't take them in.

I've mentioned the *Gorillas in Our Midst* video, in which a lady dressed in a gorilla suit walks onscreen during a basketball game pounding her chest and 50 percent of all viewers don't see her. If 50 percent of normal human beings can't see a lady dressed up like a gorilla, it's small wonder employees in meatpacking plants don't notice jiggly chains.

In their book *Inattentional Blindness*, Arien Mack at the New School for Social Research in New York City and Irvin Rock, who was a professor at the University of California, Berkeley, until he died in 1995, explain that people don't *consciously* see any object unless they are paying direct, focused attention to that object.[5] This means that a human being walking through an alley won't see, much less be bothered by, sparkling puddles or shiny spots on metal or jiggling chains. None of that stuff is there for them unless they're *looking* for it. Normal human beings are blind to anything they're not paying attention to.

My experience with animals, and with my own perceptions, is that animals and autistic people are different from normal people. Animals and autistic people don't have to be paying attention to something in order to see it. Things like jiggly chains pop out at us; they *grab* our attention whether we want them to or not.

For a normal human being, almost nothing in the environment pops. That means it's practically impossible for a human being to actually *see* something brand-new in the first place. People probably don't like novelty any more than animals do, but people don't get exposed to much novelty, because they don't notice it when it's there. *Humans are built to see what they're expecting to see,* and it's hard to *expect* to see something you've never seen. New things just don't register.

The research on inattentional blindness was shocking, because psychologists had always thought there were all kinds of things in the visual world that automatically grabbed people's attention—like an airplane blocking a runway. But it turns out that's not true. There *are* a few things that seem to grab people's attention, like the sight or sound of your own name, or large-sized objects, or—this one took me by surprise—cartoon happy faces. Not cartoon *sad* faces; a cartoon sad face is just as invisible as everything else for people who aren't actively paying attention. But a cartoon happy face will snatch people out of their inattention.

I wish they'd done some comparative research with animals and autistic people, because my guess is that animals and autistic people either don't have inattentional blindness at all, or don't have nearly as much of it as normal people do. Animals definitely act like they see everything, because you can't get anything past a cow. That's one of the reasons why a ranch owner has to correct every wrong detail, because a cow will *see* every wrong detail.

Autistic people are the same way. I know a teenage autistic boy who's a lot like those cattle trying to walk through a jiggly, sparkly chute. This boy is sixteen years old, and a couple of years ago he suddenly got focused on all the screws in the hallways at his school. He had to stop and touch each and every single one every time he went from one classroom to another. He's not scared like my cattle, but he definitely balks, and it takes forever to get him from one place to another. It's a good thing his aide has a sense of humor. The way he sees it, the boy is checking all the screws to make sure they're screwed in all the way—"He's making sure this place isn't going to fall down on top of us." He might be right about that.

I always thought the reason autistic people are so much more aware of details was that we're visual instead of verbal. I thought it was a right brain/left brain difference. For most people the left brain is verbal, the right brain is visual.

But research has found that both sides of the brain have problems in autism.[6] Based on my own experience and on my work with animals, I'm working from the hypothesis that you can understand a lot about animals and autistic people by focusing on another basic brain difference: the difference between higher parts of the brain and lower parts. The reason normal people have such a hard time seeing (and probably hearing, smelling, tasting, and feeling) details is that their *frontal lobes,* which are at the top of the brain, get in the way. Animals and autistic people see detail either because their frontal lobes are smaller and less developed (in the case of animals), or because they're not working as well as they could be (in the case of autistic people).

I'll get to that next.

LIZARD BRAINS, DOG BRAINS, AND PEOPLE BRAINS

When you compare human and animal brains, the only difference that's obvious to the naked eye is the increased size of the *neocortex* in people. (Usually the words "neocortex" and "cerebral cortex" mean the same thing, but some researchers use "neocortex" to mean the newer, six-layered part of the cerebral cortex. I'm using "neocortex" and "cerebral cortex" interchangeably.) The neocortex is the top layer of the brain, and includes the frontal lobes as well as all of the other structures where higher cognitive functions are located.

The neocortex is wrapped around all the *subcortical* or lower brain structures, which are the seat of emotions and life support functions in people and animals. In humans the neocortex is so thick compared to the lower brain structures that it's the size of a peach compared to a peach pit. In animals the cortex is much smaller. It's so small that in some animals the "peach" is the same size as the "pit"; the neocortex is the same size as all the lower brain structures.

As a general rule, the more intelligent the animal species, the bigger the neocortex. If you remove the neocortex, you can't tell an animal brain apart from a human brain, just to look at them. I had a hands-on lesson in this in grad school when I dissected a human brain and a pig brain in a class I took at the University of Illinois. The pig brain was a big shock for me, because when I compared the lower-level structures like the *amygdala* to the same structures in the human brain I couldn't see any difference at all. The pig brain and the human brain looked *exactly* alike. But when I looked at the neocortex the difference was huge. The human neocortex is visibly bigger and more folded-up than the animal's, and anyone can see it. You don't need a microscope.

Comparing animal brains to human brains tells us two things.

Number one: animals and people have different brains, so they experience the world in different ways—

and

Number two: animals and people have an awful lot in common.

To understand why animals seem so different from normal human beings, yet so familiar at the same time, you need to know that the human brain is really three different brains, each one built on top of the previous at three different times in evolutionary history. And here's the really interesting part: each one of those brains has its own kind of intelligence, its own sense of time and space, its own memory, and its own *subjectivity*. It's almost as if we have three different identities inside our heads, not just one.

The first and oldest brain, which is physically the lowest down inside the skull, is the *reptilian brain*.

The next brain, in the middle, is the *paleomammalian brain*.

The third and newest brain, highest up inside your head, is the *neomammalian brain*.

Roughly speaking, the reptilian brain corresponds to that in lizards and performs basic life support functions like breathing; the paleomammalian brain corresponds to that in mammals and handles emotion; and the neomammalian brain corresponds to that in primates—especially people—and handles reason and language. *All* animals have some neomammalian brain, but it's much larger and more important in primates and in people.

The three brains are connected by nerves, but each one has its own personality and its own control system: the "top" doesn't control the "bottom." Researchers used to think that the highest part of the brain was in charge, but they no longer believe this. That means we humans probably really do have an *animal nature* that's separate and distinct from our *human nature*. We have a separate animal nature because we have a separate animal brain inside our heads.

The reason we have three separate brains instead of just one is that evolution doesn't throw away things that work. When a structure or a protein or a gene or anything else works well, nature uses it again and again in newly evolved plants and animals. The word for this is *conservation*. Biologists say that evolution *conserves* structures that work.

Paul MacLean, the originator of the *three-brain theory*, believes that evolution simply added each newly evolved brain on top of the one that came before, without changing the older brain. He calls this the *triune brain theory*.[7]

In other words, if you're Mother Nature, and you've got a lot of lizards running around the world breathing, eating, sleeping, and waking up just fine, you don't create a whole brand-new dog breathing system when it comes time to evolve a dog. Instead, you add the new dog brain on top of the old lizard brain. The lizard brain breathes, eats, and sleeps; the dog brain forms dominance hierarchies and rears its young.

The same thing happens all over again when nature evolves a human. The human brain gets added on top of the dog brain. So you have your lizard brain to breathe and sleep, your dog brain to form wolf packs, and your human brain to write books about it. In a lot of ways evolution is like building an addition onto your house instead of tearing down the old one and building a new one from the ground up.

TRAPPED INSIDE THE BIG PICTURE

What the neocortex does better than the dog brain or the lizard brain is tie everything together. The whole neocortex is one big *association cortex*, making connections between all kinds of things that stay more separate for animals. For instance, take the fact that humans have *mixed emotions*. A human can love and hate the same person. Animals don't do that. Their emotions are simpler and cleaner, because categories like *love* and *hate* stay separate in their brains.

Another example: humans make rapid generalizations from one situation to another; animals don't. A generalization depends on making an association between one situation or object and another, similar situation or object. Compared to humans, animals generalize so little that one of the most important aspects of any animal training program is getting the animal to make a generalization from the training situation to the rest of his life. A dog can learn to perform tasks at a training school and not know how to perform them at home, because school and home are separate categories. His brain doesn't automatically associate the two. I'll talk about this more in other chapters.

Inside the neocortex, the frontal lobes, which sit behind your forehead, are the final destination for all the information that's floating around your brain. They pull everything together.

Although growing a big neocortex gave us our "book smarts," we paid a price. For one thing, bigger frontal lobes probably made humans a lot more vulnerable to brain damage and dysfunction of just about any kind. I wonder whether this explains why you don't often see animals with developmental disabilities. Estimates of the incidence of mental retardation range from 1 percent of the U.S. population up to as high as 3 percent, and it doesn't seem like there's anywhere near that level in animals. It's possible we humans don't know what a developmental disability in an animal looks like, but I also question whether animals might be less vulnerable to developmental disabilities in the first place because their frontal lobes are less developed.

Frontal lobe functions are the first to go, whether the problem is a traumatic head injury, a developmental disability, old age, or just plain lack of sleep. Worse yet, if you damage any part of your brain in an accident or a stroke you wind up with frontal lobe problems even when your frontal lobes weren't touched.

People always thought this was because the last structure to evolve is the most delicate, while the older structures have been around so long they've become incredibly robust. But a neuropsychologist named Elkhonon Goldberg at New York University School of Medicine, who wrote a fantastic book about frontal lobe functions called *The Executive Brain,* has a different theory. He thinks that while the frontal lobes may be more fragile, there is another factor involved, which is that every other part of the brain is connected to them. When you damage *any* part of the brain, you change input to the frontal lobes, and when you change input, you change output. If the frontal lobes aren't getting the right input, they don't produce the right output even though structurally they're fine. So all brain damage ends up looking like frontal lobe damage, whether the frontal lobes were injured or not.[8]

I think he's right about this, because frontal lobe problems are a big part of autism, and our frontal lobes are structurally pretty good. A major autism researcher told a journalist friend of mine that if you compared the brain scan of an autistic child to the scan of a sixty-year-old CEO, the autistic child's brain would look better. In other words, the normal brain shrinkage people experience with age makes your brain look more "abnormal" than autism does. There *are* some structural differences between autistic brains and normal brains, but they're so small you can't see them on a regular MRI, and probably every person has structural brain differences to that degree.

Of course, the fact that a brain difference is tiny doesn't mean its effect is tiny. The researcher also said that a brain difference could be subtle but significant. But he added that there's nothing about the anatomy of the autistic brain that told him autism can't eventually be treated by medication the same way psychiatric disorders can be treated.

Until we learn more, I am assuming that one of the problems in autism isn't bad frontal lobes; it's *bad input* into the frontal lobes.

Bad input can happen to normal people, too. Just being incredibly tired and sleep-deprived will lower your frontal lobe function, and the aging process hurts the frontal lobes much more than any other part of the brain.

That brings me back to animals. The good news is: when your frontal lobes are down, you have your animal brain to fall back on. That's exactly what happens, too. *The animal brain is the default position for people.* That's why animals seem so much like people in so many ways: they *are* like people. And people are like animals, especially when their frontal lobes aren't working up to par.

I think that's also the reason for the special connection autistic people like me have to animals. Autistic people's frontal lobes almost never work as well as normal people's do, so our brain function ends up being somewhere in between human and animal. We use our animal brains more than normal people do, because we have to. We don't have any choice. *Autistic people are closer to animals than normal people are.*

The price human beings pay for having such big, fat frontal lobes is that normal people become oblivious in a way animals and autistic people aren't. Normal people stop *seeing* the details that make up the big picture and see only the big picture instead. That's what your frontal lobes do for you: they give you the big picture. Animals see all the tiny little details that go into the picture.

EXTREME PERCEPTION: THE MYSTERY OF JANE'S CAT

Compared to humans, animals have astonishing abilities to perceive things in the world. They have *extreme perception*. Their sensory worlds are so much richer than ours it's almost as if we're deaf and blind.

That's probably why a lot of people think animals have ESP. Animals have such incredible abilities to perceive things we can't that the only explanation we can come up with is extrasensory perception. There's even a scientist in England who's written books about animals having ESP. But they don't have ESP, they just have a supersensitive sensory apparatus.

Take the cat who knows when its owner is coming home.[9] My friend Jane, who lives in a city apartment, has a cat who always knows when she's on her way home. Jane's husband works at home, and five minutes before Jane comes home he'll see the cat go to the door, sit down, and wait. Since Jane doesn't come home at the same time every day, the cat isn't going by its sense of time, although animals also have an incredible sense of time. Sigmund Freud used to have his dog with him every time he saw a patient, and he never had to look at his watch to tell when the session was over. The dog always let him know. Parents tell me autistic kids do the same thing. The only explanation Jane and her husband could come up with was ESP. The cat must have been picking up Jane's I'm-coming-home-now thoughts.

Jane asked me to figure out how her cat could predict her arrival. Since I've never seen Jane's apartment I used my mother's New York City apartment as a model for solving the mystery. In my imagination I watched my mother's gray Persian cat walk around the apartment and look out the window. Possibly the cat could see Jane walking down the street. Even though he would not be able to see Jane's face from the twelfth floor he would probably be able to recognize her body language. Animals are very sensitive to body language. The cat would probably be able to recognize Jane's walk.

Next I thought about sound cues. Since I am a visual thinker I used "videos" in my imagination to move the cat around in the apartment to determine how it could be getting sound cues that Jane would be arriving a few minutes later. In my mind's eye I positioned the cat with its ear next to the crack between the door and the door frame. I thought maybe he could hear Jane's voice on the elevator. But as I played a tape of my mother getting onto the elevator in the lobby, I realized that there would be many days when Mother would ride the elevator alone and silent. She would speak on the elevator for only some of the trips—when there were other people in the elevator car with her—but not all of them.

So I asked Jane, "Is the cat always at the door, or is he at the door only sometimes?"

She said the cat is always at the door.

That meant the cat had to be hearing Jane's voice on the elevator every day. After I questioned her some more, Jane finally gave me the crucial piece of information that solved the cat mystery: her building does not have a push-button elevator. The elevator is operated by a person. So when Jane got on the elevator she probably said "Hi" to the operator.

A new image flashed into my head. I created an elevator with an operator for my mother's building. To make the image I used the same method people use in computer graphics. I pulled an image of my mother's elevator out of memory and combined it with an image of the elevator operator I saw one time at the Ritz in Boston. He had white gloves and a black tuxedo. I lifted the brass elevator control panel and its tuxedoed operator from my Ritz memory file and placed them inside my mother's elevator.

That was the answer. The fact that Jane's building had an elevator operator provided the cat with the sound of Jane's voice while Jane was still down on the first floor. That's why the cat went to the door to wait. The cat wasn't *predicting* Jane's arrival; for the cat Jane was already home.

DIFFERENT SENSE ORGANS

Cats have really good hearing, so Jane's cat was using a sensory capacity we humans don't have. Animals have all kinds of sensory abilities we don't have, and vice versa. (Our color vision is a good example of a sensory capacity we have that a lot of animals don't.) Dogs can hear dog whistles; bats and dolphins can use sonar to "see" a moving object at a distance (a flying bat can actually spot and classify a flying beetle from thirty feet away); dung beetles can perceive the polarization of moonlight. I know dung beetles are insects, not animals, but an insect's brain is so tiny it makes the things their sensory system can handle even more miraculous.

There are two things going on with extreme perception in animals: one is the different set of sense *organs* animals have, and the other is a different way of *processing* sense data in the brain. With Jane's cat, I'm talking mostly about a different physical capacity to hear sounds humans can't.

There are hundreds or maybe even thousands of examples of this in the animal world, lots of which we probably still don't know about. A good example is the *silent thunder* of elephants. It wasn't until the 1980s that a researcher named Katy Payne, of Cornell University, figured out that elephants communicate with one another using infrasonic sound waves too low for humans to hear.[10] People who studied elephants had always wondered how elephant families managed to coordinate their movements with family members miles away. An elephant family could be split up for *weeks,* and then meet up at the same place at the same time. They had to be communicating with one another somehow, but they were way out of the range any human could either see or shout across.

Katy Payne made a lucky guess about infrasonic sound when she felt "a throbbing in the air" next to the elephant cages at the Portland Zoo in Oregon. She'd had the same feeling as a child when the organ played in church. She started to think maybe the elephants were communicating with each other in a super-low range humans don't hear. That would solve the problem of the long-distance communication, because infrasonic sound travels a lot farther than sound waves in the register humans do hear.

She turned out to be right. Elephants "roar" out to each other below our level of hearing. During the daytime an elephant can hear another elephant calling him from at least as far away as two and a half miles. At nighttime, because of temperature inversions, that distance can go up by an order of magnitude to as much as twenty-five miles. It's a huge distance.

Now it turns out that elephants may be talking to one another through the ground, not just the air. Caitlin O'Connell-Rodwell, a biologist at Stanford, is working on this. She believes elephants can probably use *seismic communication*—making the ground rumble by stomping on it—to communicate with other elephants as far away as twenty miles.

She figured this out by watching the elephants in the Etosha National Park in Namibia. Right before another herd of elephants arrived, the elephants she was watching would start to "pay a lot of attention to the ground with their feet."[11] They'd do things like shift their weight or lean forward, or lift a foot off the ground. They were listening.

Dr. O'Connell-Rodwell thinks the animals are probably using the pads of their feet like the head of a drum. She and her team are also dissecting elephant feet to see whether they have *pascinian* and *meissner corpuscles*, which are special sensors elephants have in their trunks to detect vibrations. If they find them in the feet, too, that's pretty good evidence elephants use seismic waves to communicate. A lot of animals communicate by thumping on the ground, including skunks and rabbits, so it won't surprise me if we find out elephants are talking to one another that way.

If elephants do have special corpuscles to detect vibrations that would be an example of an animal species having extreme perception because they're built differently and have different sense organs. Animals have all kinds of sense receptors we don't. Another example: dolphins have an oil-filled sac in their foreheads, underneath their forehead bumps, that they use for sonar. The dolphin sends a sound through the oil (which "focuses" the sound) and out to objects in the water. The sound bounces back to the dolphin and his brain forms a sound picture of what's out there. Humans can't use sonar because humans don't have any of the necessary sense structures.

Humans also have sensory receptors animals don't, like the huge number of cones in our retina for seeing color.

I've been talking mostly about vision, but all the other senses are different in different animals, too. There's some fascinating new research about the relationship between vision and smell in New World versus Old World primates. Old World primates are the famous ones everyone knows about: gorillas, chimpanzees, baboons, orangutans, macaques, humans. New World primates are the smaller animals we call monkeys. New World primates usually live in trees in Central and South America; they have long prehensile tails and flat noses. Tamarins, squirrel monkeys, sakis, and marmosets are all New World monkeys.

Old World primates, like baboons, chimpanzees, and macaques, have trichromatic, three-color vision, but most of the New World monkeys (spider monkeys, marmosets, capuchins) only have dichromatic, two-color vision. (Some New World females have trichromatic vision, but not all.)

What's interesting about this is that Old World primates and humans also have very poor ability to smell pheromones, which are chemical signals animals emit as a form of communication. (Most peo-

ple think of pheromones as sexual signals, like the pheromones a female in heat emits, but a pheromone is *any* chemical used for communication. Ants, for instance, leave trails of scents behind them for
other ants to follow.) About a year ago researchers found that Old
World primates and humans both have so many mutations in a gene
called TRP2, which is part of the *pheromone signaling pathway*, that
it's not working anymore. In the course of evolution, the pheromone
system in Old World primates, including humans, broke down.

It turns out that when we gained three-color vision we probably
lost pheromone signaling. Jianzhi George Zhang, an evolutionary
biologist at the University of Michigan, ran a computer simulation
to find out *when* the TRP2 gene started to deteriorate, and discovered that TRP2 went into decline at the same time Old World primates were developing trichromatic color vision, around 23 million
years ago.[12]

Probably what happened was that once Old World primates could
see in three colors they started using their vision to find a mate, instead
of their sense of smell. That theory fits with the fact that a lot of Old
World primate females have bright red sexual swellings when they're
fertile, while New World monkeys do not. Once monkeys no longer
needed a good sense of smell to reproduce successfully, their ability to
smell probably went into decline as a direct result.

That would have happened because *use it or lose it* is a principle in
evolution. If monkeys with a poor sense of smell can reproduce just
as well as monkeys with an excellent sense of smell, the monkeys
with the poor ability pass all of their weak or defective smell genes
on to their offspring, and any spontaneous new mutations in the
smell genes don't get winnowed out. It looks like that's what happened to Old World primates. The normal mutations that happen in
the process of reproduction just kept accumulating until no primates
had a working copy of TRP2 anymore. Improved vision came at a
cost to their sense of smell.

SAME BRAIN CELLS, DIFFERENT PROCESSING

So far I've been talking about the sense organ or sense receptor part
of animal perception: animals have different sensory organs than we

do, organs that let them see, hear, and smell things we can't. But the other half of the story is where things get interesting, and that is the differences in brain processing.

All sensory data, in any creature, has to be processed by the brain. And when you get down to the level of brain cells, or neurons, humans have the same neurons animals do. We're using them differently, but the cells are the same.

That means that theoretically we *could* have extreme perceptions the way animals do if we figured out how to use the sensory processing cells in our brains the way animals do. I think this is more than a theory; I think there are people who already do use their sense neurons the way animals do. My student Holly, who is severely dyslexic, has such acute auditory perception that she can actually hear radios that aren't turned on. All appliances that are plugged in continue to draw power, even when they're turned off. Holly can hear the tiny little transmissions a turned-off radio is receiving. She'll say, "NPR is doing a show on lions," and we'll turn the radio on and sure enough: NPR is doing a show on lions. Holly can hear it. She can hear the hum of electric wires in the wall. And she's incredible with animals. She can tell what they're feeling from the tiniest variations in their breathing; she can hear changes the rest of us can't.

Autistic people almost always have excruciating sound sensitivities. The only way I can describe how a lot of sounds affect me is to compare it to staring straight into the sun. I get overwhelmed by normal sounds in the environment, and it's painful. Most autism professionals talk about this as just being super-*sensitive*, which is true as far as it goes. But I think autistic people are also super-*perceptive*. They're hearing things normal people aren't, like a piece of candy being unwrapped in the next room.

It happens with vision, too; a lot of autistic people have told me they can *see* the flicker in fluorescent lighting. Holly's the same way. She can barely function in fluorescent lighting because of it. Our whole environment is built to the specifications and limitations of a normal human perceptual system—and that's not the same thing as a normal animal perceptual system, or as a *normal-abnormal* human system like a dyslexic person's system, or an autistic person's. There are probably huge numbers of people who don't fit the normal environment. Even worse, half the time they probably don't even realize they don't fit, because this is the only environment they've ever been in, so they don't have a point of comparison.

Some researchers say that people like Holly have *developed* super-sensitive hearing because their visual processing is so scrambled. Super-sensitive hearing is a compensation, in other words. That's always the explanation researchers give for the super-hearing of blind people; people who are blind have built up their hearing to compensate for not being able to see.

I'm sure that's true, but I don't think it's the whole story. I think the potential to be able to hear the radio when it's turned off is *already there* inside everyone's brains; we just can't access it. Somehow a person with sensory problems figures out how to get to it.

I have two reasons for thinking this. First, there are a lot of cases in the literature of people suddenly developing extreme perception after a head injury. In *The Man Who Mistook His Wife for a Hat* Oliver Sacks has a story about a medical student who was taking a lot of recreational drugs (mostly amphetamines). One night he dreamed that he was a dog. When he woke up he found that all of a sudden, literally overnight, he had developed super-heightened perceptions, including a heightened sense of smell. When he went to his clinic, he recognized all twenty of his patients, before he saw them, *purely by smell*. He said he could smell their emotions, too, which is something people have always suspected dogs can do. He could recognize every single street and shop in New York City by smell, and he felt a strong impulse to sniff and touch things.[13]

His color perception was much more vivid, too. All of a sudden he could see dozens of shades of colors he'd never seen before— dozens of shades of the color brown, for instance.

This happened overnight. It's not like he lost some other sense and then built up his sense of smell over time to compensate. He dreamed he was a dog and the next morning woke up able to smell things like a dog. The actor Christopher Reeve had a similar experience right after his accident. All of a sudden he had an incredibly heightened sense of smell.

The other important thing to know about this guy is that he hadn't had any big brain injury that anyone knew about. Dr. Sacks assumes that the heavy drug usage was probably the cause, but there's no way of knowing. The student continued to function in medical school just fine, and his vision and sense of smell went back to normal about three weeks later. Of course, some part of his brain could have been temporarily incapacitated, but if it was, there's no obvious way that being able to smell people the way a dog smells people helped him compensate for whatever might have been wrong. The most likely explanation is that he always had an ability to smell like a dog and see fifty different shades of brown, but he just didn't know it and couldn't access it. Somehow his heavy amphetamine usage must have opened up the door to that part of his brain.

My other reason for thinking everyone has the potential for extreme perception is the fact that animals have extreme perception, and people have animal brains. People use their animal brains all day long, but the difference is that people *aren't conscious of what's in them*. We'll talk about this in the last chapter. A lot of what animals see normal people see, too, but normal people don't know they're seeing it. Instead, a normal person's brain uses the detailed raw data of the world to form a generalized concept or schema, and that's what reaches consciousness. Fifty shades of brown turn into just one unified color: brown. That's why normal people see only what they expect to see—because they can't *consciously experience* the raw data, only the schema their brains create out of the raw data.

Normal people see and hear schemas, not raw sensory data.

I can't prove that humans are taking in the same things animals are, but we do have proof that humans are taking in way more sensory data than they realize. That's one of the major findings of the inattentional blindness research. It's not that normal people don't see the lady dressed in a gorilla suit at all; it's that their brains screen her out before she reaches consciousness.

We know people see things they don't know they see because of years of research into areas like *implicit cognition* and *subliminal perception*. Dr. Mack and Dr. Rock, who wrote *Inattentional Blindness*, adapted some of these studies for their inattentional blindness research. They'd do things like ask their subjects to tell them which arm of a cross that flashed onto a computer screen for about 200 milliseconds was longer. Then, on some of the trials, there'd be a word like "grace" or "flake" printed on the screen, too. Most people didn't notice the word. They were paying attention to the cross, so they didn't see it.[14]

But Dr. Rock and Dr. Mack showed that many of them *had* seen the words unconsciously. Later on, when they gave subjects just the first three letters of the word—*gra* or *fla*—and asked them to finish them with any word that came to mind, 36 percent answered "grace" or "flake." Only 4 percent of the control subjects—these were people who hadn't been subliminally exposed to any words at all—came up with "grace" or "flake." That's a huge difference and can only mean that the subjects who were subliminally exposed to "grace" and "flake" really did see "grace" and "flake." They just didn't know it.

So we know that people perceive lots more than they realize consciously. Drs. Rock and Mack say that inattentional blindness works at a *high level of mental processing,* meaning that your brain does a lot of processing before it allows something into consciousness. In a normal human brain sensory data comes in, your brain figures out what it is, and only then does it decide whether to tell you about it, depending on how important it is. A lot of processing has already taken place before a normal human becomes conscious of something in the environment. (Drs. Rock and Mack use the phrase *high level* to mean advanced *processing,* not necessarily higher levels of the brain. They don't discuss neuropsychology, just cognitive psychology.)

There are a few things that always *do* break through to consciousness. I mentioned that people almost always notice their names in the middle of a page of text no matter how hard they're concentrating on something else; they will also notice a cartoon smiley face. But if you change the face just a tiny bit—turn the smile upside down so it's a frown, for instance—people don't see it. This is more evidence for the fact that your brain thoroughly processes sensory data before allowing it to become conscious. With the smiley face your brain has to have processed it to the level of knowing it's a face and even that it's a *smiling* face before it lets the face into conscious perception. Otherwise you'd see the frowny face as often as you saw the smiley face. It's the same principle with your name. If your name is "Jack," the word "Jack" will pop out at you in the middle of a page. But the letters "Jick" won't. That means your brain processes the word "Jack" all the way up to the level of knowing that it's your name before your brain admits "Jack" into consciousness.

We don't know why humans have inattentional blindness. Maybe inattentional blindness is the brain's way of filtering out distractions. If you're trying to watch a basketball game and a lady gorilla comes into view, your brain screens her out because she's not supposed to be there, and she's not relevant to what you're trying to do, which is watch a game. Your nonconscious brain takes a look at the lady gorilla and decides she's a distraction.

Being able to filter out distractions is a good thing; just ask anyone who can't filter things out, like a person with *attention deficit hyperactivity disorder.* It's hard for humans to function intellectually when every little sensory detail in their environment keeps hijacking their attention. You go into information overload.

But humans probably paid a price for developing the ability to filter out ladies wearing gorilla suits, which is that normal people can't *not* filter out distractions. A normal brain automatically filters out irrelevant details, whether you want it to or not. You can't just tell your brain: be sure and let me know if anything out of the ordinary pops up. It doesn't work that way.

Autistic people and animals are different: we can't filter stuff out. All the zillions and zillions of sensory details in the world come into our conscious awareness, and we get overwhelmed. There's no way to know exactly *how* close an autistic person's sensory perceptions are to an animal's. There are probably some big differences, if only for the reason that animal perceptions are normal for animals, while autistic people's perceptions are *not* normal for people.

But I think many or even most autistic people experience the world a lot the way animals experience the world: as a swirling mass of tiny details. We're seeing, hearing, and feeling all the things no one else can.

3. Animal Feelings

RAPIST ROOSTERS

We've been doing some strange things to animals' emotional makeup in our breeding programs. When I was just starting my work with chickens a few years ago, I visited a chicken farm. Inside the barn where all the chickens lived I found a dead hen lying there on the floor. She was all cut up, and her body was fresh. I was horrified.

I went back to the farmer and I said, *"What* was *that?"*

He told me the rooster did it: the rooster killed the hen. He acted like that was a perfectly normal thing for a rooster to do. He wasn't happy that his roosters were killing his hens; he just thought that's the way it was.

I knew that couldn't be right. If roosters killed hens in nature, there wouldn't be any chickens. But people raising animals in captivity tend to forget this basic fact of life. A lady who raises llamas told me recently that one of her males had tried to bite the testicles off another male. I told her that's definitely not normal. If llamas bit off each other's testicles in the wild there wouldn't be any llamas.

The chicken farmer told me that *half* of his roosters were rapist-murderers. I was stunned when I heard that. There is no species alive in nature where half the males kill reproductive-age females. There had to be something seriously wrong with those birds.

So when I got home I immediately talked to one of my students whose family were backyard breeders. They had a small side business raising and breeding chickens in the backyard. She'd never even heard of a rooster killing a hen. Then I called my good friend Tina Widowski, who was a specialist in chickens, and she said absolutely not: normal roosters do not kill hens.

Tina knew about the killer roosters, and she told me what the deal was. Ian Duncan from the University of Guelph in Canada had studied the roosters and had found that the rooster courtship program had gotten accidentally deleted in about half of the birds. A normal rooster does a little courtship dance before trying to mate with a hen. The dance is hardwired into the rooster's brain; it is instinctual behavior, or what animal ethologists call a *fixed action pattern*. All normal roosters do it.

The dance triggers a fixed action pattern in the hen's brain, and she crouches down into a sexually receptive position so the rooster can mount her. *She doesn't crouch down unless she sees the dance.* That's the way her brain is wired.

But half of the roosters had stopped doing the dance, which meant that the hens had stopped crouching down for them. So the roosters had become rapists. They jumped on the hens and tried to mate them by force, and when the hen tried to get away, the rooster would attack her with his spurs or his toes and slash her to death.

SINGLE-TRAIT BREEDING

The rapist roosters were a side effect of *single-trait breeding,* which is selectively breeding animals to increase or decrease just one or two desirable traits, like fast growth (to decrease feed costs and time to market) or heavy muscling (to increase the amount of meat per bird). The breeders focus totally on those traits and nothing else.

Single-trait breeding isn't quite as simple as just mating fast-growing, big-muscled males to fast-growing, big-muscled females, because when you do that fertility tends to go out the window. You get animals who have trouble reproducing themselves. So breeders mate females who have fast growth, big muscles, *and* sound fertility to males who just have fast growth and big muscles. They don't worry about fertility levels in the males, because even a low-fertility male can still fertilize the eggs of a female with good fertility. You end up with a *hybrid* chicken, which means a cross between two different lines. All of the chicken meat and eggs we eat come from hybrid birds.

Single-trait breeding works fast in chickens because their reproductive cycle is so short. A chicken egg has to be incubated for only twenty-one days before it hatches, and a newly hatched female chick will be ready to lay fertile eggs five months later. That's two generations of chickens in one year, which means that in three to five years the genetic line can be changed completely.

The problem with single-trait breeding is that when you breed for one trait you end up changing other traits, too: there are always unintended consequences. That's what happened with the roosters.

The rapist roosters didn't come along until breeders had gone through at least three different single-trait breeding programs over a number of years. The first goal the industry pursued was to develop faster-growing chickens who could be taken to market sooner. They mated the fastest-growing hens with the fastest-growing roosters and voilà: faster-growing chickens. In practice it's more complicated than this, because they also do various genetic calculations, but the basic approach is to breed *fast* to *fast*.

Like every single-trait breeding program, this one had some unintended consequences although they weren't as severe as the rapist roosters. Mainly the faster-growing chickens tended to have weaker legs and hearts. The weak hearts meant a higher incidence of *flip-over disease,* which is a nice way of saying that the chickens' hearts gave out. Heart failure in chickens got the name flip-over disease because that's what it looks like. When a chicken has heart failure it suddenly flips over and dies.

The next goal was to breed chickens with bigger breasts because people like to eat white meat. That program was successful, too; they got chickens with bigger breasts. This time they got a lot more problems, though, because the chickens grew so big that their legs couldn't handle their weight. Many of the chickens were so lame they couldn't walk to the feeder, and some of their legs were deformed and bent, with fluid-filled swellings.

The chickens were probably in constant pain. One study found that the lame chickens would choose to eat bad-tasting feed laced with painkillers over normal-tasting feed, which is good evidence that the chickens were suffering.

They also had a much higher rate of flip-over disease, because their hearts were too small to pump blood to their huge bodies. It was like trying to run a Mack truck on a Volkswagen engine, and their hearts gave out.

The big-breasted chickens were a disaster. No one wants to breed chickens who are lame and in pain, but even if you didn't care about the chickens' well-being, no business can sell deformed drumsticks to the food market. They had to do something, so they started breeding for strength and *livability*, which means the chickens' overall health and ability to grow and thrive instead of dying young.

They didn't just go backward a step to the earlier chickens with the smaller breasts, because they wanted to have their cake and eat it, too. They wanted fast-growing, big-breasted chickens with strong legs and sound hearts. Breeder colonies think like software companies. If there are problems with Chicken 3.2 they don't go back to Chicken 3.1. They go forward to 3.3.

So in another few years they had big, strong chickens with big, heavy thick legs and stronger hearts and it looked like they had their dream chicken. Then nature threw them a curveball and they got raping, murdering roosters. No one has any idea why or how the genes for hearts and bones are related to the genes for mating behavior, but obviously they are.

This kind of thing happens all the time when breeders over-select for a single trait. You get warped evolution.

The really bad thing was that the change happened slowly enough that the farmers and probably the breeder colonies, too, didn't realize they'd created a monster. Nobody noticed what was happening. As the roosters got more and more aggressive, the humans unconsciously adjusted their perceptions of how a normal rooster should act. It was a case of *the bad becoming normal,* and it's a big danger in selective breeding programs. I've seen it many times.

SELECTION PRESSURE

Human beings are constantly changing the *selection pressure* on animals, whether we want to or not. Selection pressure is the term for what happens when the environment influences or *selects* which members of a species live long enough to reproduce successfully and which do not. Selection pressure can help new traits get stronger and more widespread in a species, and the absence of selection pressure can cause long-standing traits to weaken or disappear.

That's what happened with the Old World primates. They probably experienced some kind of random genetic mutation that gave them better color vision; then their improved color vision was so useful for finding food that the animals with the best color vision also had the best chance of staying alive long enough to reproduce. After their babies were born, of course, the trichromatic vision animals would also have had the best chance of finding enough food to feed their offspring, so the babies—many or most of whom would have inherited the new trichromatic vision mutation—also lived to adulthood and reproduced. That's how selection pressure strengthens a trait; it gives animals who have the trait a reproductive advantage. Over succeeding generations their genes spread through the population.

At the same time, once vision became *more* important for finding food, smell probably became *less* important. This would have meant that animals with an excellent sense of smell were no more likely to have babies than animals with a poor sense of smell: everyone could reproduce no matter how good or bad their sense of smell. The selection pressure for good smell genes was gone.

As a direct result, Old World primates' sense of smell would have gone into decline. Genetic mutations happen all the time when animals reproduce, due to simple copying errors in the DNA. Some mutations are good, some are bad, and some don't make a difference one way or another. Selection pressure saves and promotes the good mutations and weeds out the bad ones. Once selection pressure is gone, a trait will go into decline through the natural process of routine genetic mutations piling up on each other until the trait or traits those genes influence has weakened or even disappeared.

THE BAD BECOMES NORMAL

When it comes to domestic animals, *we're the environment*. We create the selection pressures. If you're a chicken breeder and you allow only the fastest-growing roosters and hens to breed you're exerting a selection pressure in favor of faster-growing chickens.

That's an example of an intentional selection pressure, but people create unconscious, accidental selection pressures all the time. For instance, you've probably heard doctors complain about patients not finishing their full course of antibiotics. The reason you're supposed to finish your antibiotics is that when you don't, you're unintentionally creating a selection pressure that favors the development of antibiotic-resistant strains. That's because you've used a half-course of antibiotics to kill off the weaker bacteria, leaving only the stronger bacteria alive to reproduce. Over several generations, that's how you get bugs that antibiotics can't kill.

With domestic animals, we are the main engine of evolution. We're constantly changing their bodies and their emotions, and it happens a lot faster than we realize. The most interesting study of this was one where the researchers gave rats from the same genetic strain to two different labs and then left them there for five years while the labs did all their normal experiments. Both of the labs bred successor generations from the original rats, which meant that researchers could compare the descendants.

At the end of five years, the researchers tested the rats and found that the two groups of descendants had developed completely different levels of natural fear. They tested this by doing *open arena* studies with each rat. In an open arena study you put one rat alone inside a well-lit open space about the size of a large tabletop, and watch to see how much exploring he does. Being prey animals, mice and rats don't like open, well-lit spaces. Only the boldest rat will do much exploring out in the open; most rats hang back at the sides of the arena and stay still.

The ancestor rats had all started out with exactly the same levels of exploratory behavior. But at the end of just five years, one group of descendant rats was much more fearful than the other.

The interesting thing was that none of the people at the labs had any idea their rats had changed, and none of them had *tried* to breed rats with different levels of fear. The two groups of rats just naturally evolved away from each other in response to different conditions at the two labs. This is what happens in unconscious selective breeding.

The researchers don't know why the rats evolved away from each other. They just know that they did. No one at either of the labs had any kind of agenda other than just to use the rats in normal psych lab studies, and there wasn't any big difference in the kinds of studies the two labs were doing that could explain why the two descendant groups diverged in personality.

My guess is that the employees at the two labs probably responded differently to aggressive behaviors without realizing it. Say I'm the lady who takes care of the rats at the first lab, and I have a couple of rats who bite—I just get rid of them, because I don't like them. At the other colony you've got the same number of rats who bite, because the rats are all from the same line. But maybe in the second lab you've got a guy with big gloves and he's kind of macho so he *doesn't* get rid of them. Rats who bite get culled out of the gene pool in the first lab, but they live and reproduce in the second lab.

That would make a difference in the open arena test because fear and aggression are related. Most of the time, the more fearful an animal the less aggressive he is, because a high-fear animal is afraid to get into a fight. High-fear animals can also be more aggressive under certain circumstances, which I'll get to in a second. But overall, fear inhibits aggression. *Under most circumstances,* rats who bite are probably less fearful, so at the first lab, the one with the hypothetical lady culling the biters, I'm selectively breeding for fearful rats, and that's what I'm getting. Rats with higher fear.

That's one possibility.

On the other hand, say it was the macho guy who was culling the aggressive rats, not the lady. Maybe he was handling his rats roughly and scaring them. If that were the case, the rats doing the biting would be the high-fear rats, not the low-fear ones. Low-fear animals are more aggressive when fighting another animal, but high-fear animals are more likely to panic and bite when they are handled roughly by humans. If the man got rid of the biting rats he would be the one changing the gene pool by selecting for calm, low-fear animals who can take a lot of rough handling. Either way you get the same result: one lab is inadvertently breeding a different kind of rat—a more confident rat with *lower* fear, in this case.

We'll probably never know what happened, but the bottom line is that some sort of unconscious, unintentional selection pressure had to have been put on one or both lines of those rats to cause them to diverge so dramatically in just five years.

That's not a bad thing on the face of it necessarily, because accidental selection pressure is *probably* less dangerous to the animal, although that's never been studied. At least with unconscious selective breeding the humans influencing evolution aren't consciously trying to change *just one aspect of the animal.* They may be unconsciously shaping a cluster of related behaviors, or they may just be less intense about changing the one behavior (like biting) that's bothering them. The lady at lab number one might not be culling every single rat who shows the tiniest bit of aggression, so the gene pool doesn't get quite so distorted as it does in formal selective breeding programs.

It's when you consciously and purposely breed animals to change one defined physical trait dramatically from what nature intended that you can definitely end up with some major emotional and behavioral problems. Moreover, when you're trying to change a physical trait you very, very often end up changing an emotional and behavioral trait, too. The body and the brain aren't two different things, controlled by two completely different sets of genes. Many of the same chemicals that work in your heart and organs also work in your brain, and many genes do one thing in your body and another thing in your brain. So if you change a gene in order to change a chicken's breast size, you're also going to change whatever that gene might have been doing in the chicken's brain, assuming you're modifying a gene that is active in both places.

This is a very serious problem in the selective breeding of animals. Over the years I have learned that when you over-select for any trait at all, eventually you get neurological damage, and neurological damage almost always means emotional damage, or at least important emotional changes. The distressing thing is that with single-trait breeding for a physical trait, nobody notices the emotional changes that are emerging right along with the altered physical trait, because *nobody is expecting to see any emotional changes.* The breeders are monitoring physical changes, not looking for emotional or behavioral changes. So breeders don't perceive how much the animals are changing emotionally until the behaviors have gotten so extreme that the alarm bells finally go off. Then they've got another big problem to fix.

With the rapist roosters, the good news is that I think they probably are getting the problem fixed now. I saw some of these chickens just a few months ago, and they all behaved just as nicely as can be. I think probably the companies are culling the rapists, but I don't know for sure, since they don't write up and publish what they're doing.

PSYCHO HENS

There was another really bad case of warped evolution with egg-laying hens up in Canada. White chickens are much more hyperactive and frantic than brown chickens, who are calm and laid-back. However, white chickens have a big advantage over brown chickens, which is that they need a lot less food to lay the same number of eggs. That's called *feed conversion*. The farm I visited wanted to breed brown chickens who could lay more eggs on less food, so they crossed them with the more feed-efficient white chickens. (They didn't want to just switch over to white chickens, because a lot of people like to eat brown eggs and only brown chickens produce brown eggs.)

In the succeeding generations some chickens were still almost solid brown, some had brown and white feathers mixed together like a tweed, and others were mostly white. The brown chickens had mature feathers, but the feathers on the white chickens were immature, soft, and runty. The central spine of each feather was short, soft, and limp, and the feather *barbs* (that's the feathery stuff that grows out of the shaft) were super-soft, almost like down.

Emotionally, the brown hens were the calmest birds, the brown and white hens were nervous, and the white hens were extremely anxious and agitated. When you walked into the barn they would go berserk squawking and hopping up and down. They were extremely hyperactive and frantic-acting. Then when they got old they'd start beating their own feathers off against the sides of their cages, until they were half nude. They were violent, too, and would peck and kill each other if they had a chance.

But nobody did anything about it because it was another case of the change happening slowly enough that the humans adjusted to the new chicken reality and perceived it as natural. The bad became normal. Finally one farmer bought some brown Hutterite chickens that they housed close to the white chickens, and the difference jumped out at them. The brown chickens could lay as many eggs as the white chickens, although they needed 10 percent more feed, but they were calm birds who didn't show any signs of agitation or anxiety. When they got old they still had all their feathers.

In the hens' case I think the reason for their psychological problems is less mysterious than whatever caused the roosters to become violent rapists. *Pure* white animals (and people) have more neurological problems than dark-skinned or dark-furred animals, because melanin, the chemical that gives skin its color, is also found in the midbrain, where it may have a protective effect.[1] You see all kinds of problems in white animals. Dalmatians with the highest ratio of white fur to black are getting close to true albinism. They're more likely to be deaf than other dogs, and they are often airheads you can't train. Black-and-white paint horses can have problems, too. It's not unusual for a paint horse to be plain crazy, especially if he has blue eyes.

You see quite a few problems in animals with blue eyes. I met a paint horse once who had one brown eye and one blue eye, and he clearly had a horse version of Tourette's. Every sixty seconds his whole body would flinch uncontrollably. And it's fairly well known that if you mate two blue-eyed huskies the offspring can have problems.

The color of the animal's skin is more important than the color of its fur. If its skin is dark, that's good. The inside of a dog's mouth should be mostly black, with some white.

True albino animals are much worse. A study by Donnell Creel, research professor of ophthalmology and visual sciences at the University of Utah, looked at all the problems and differences in albino animals, and concluded that researchers should not be using albino animals in their research, because albino animals are not normal. Albino animals like the white laboratory rats people used for years probably aren't even good for drug research, because melanin binds to some of the chemicals used in medications, so an albino animal's response to a medication can be completely different from a non-albino animal's.[2]

In the wild there are very few all-white animals apart from polar bears and the occasional white wolf. But polar bears and white wolves both have dark skin; it's just their fur that's white. They aren't albino. It's when the skin is all pink or all white that the animal has problems. Albino animals are occasionally born in nature, but their survival rate is low because of all their problems. I am definitely against humans doing things like deliberately breeding albino Doberman pinschers because they look so pretty. These animals are not normal, and they suffer. People who own albino Dobermans report that their dogs have poor vision, intolerance to sunlight, skin lesions, and problems with temperament, usually aggression. In one survey 11 percent of owners said their dogs had bitten people.[3] That's an enormously high number considering how rare dog bites are in comparison to the number of dogs living with humans.

So it's not surprising that an all-white chicken would have a lot of emotional problems, though we don't know why breeding a chicken to need less food would turn her feathers and skin white in the first place. How is feed conversion related to feather color? We don't know, and we could probably learn a lot about the biology of emotions if we studied the unintended behavioral changes that have come out of these selective breeding programs.

Whenever I talk about white animals in my lectures, people want to know whether what I'm saying applies to white and black people. The answer is that it doesn't, because white people aren't really white. White humans have melanin in their skin, and get tan when they spend time in the sun. Caucasians' skin color evolved the same way everyone else's skin color evolved, through the forces of natural selection operating without conscious human interference. That's why you see emotional and behavioral differences in all-white animals that you don't see in "all-white" people.

Nature doesn't evolve a Dalmatian. The Dalmatian has been artificially bred to be mostly white, and is starting to be closer to albinos than to normally pigmented animals. It's not an albino, but it's getting there.

HOW PEOPLE CHANGE ANIMALS' EMOTIONS

So far I've been talking mostly about accidental changes that come
out of single-trait breeding programs, but people also change animals all the time in a much more natural way. That happens because
people are in charge of domestic animals' lives, and we make or
influence the decisions about which animals get to reproduce and
which animals don't.

A lot of the time—maybe even most of the time—these changes
aren't bad at all. Some of the changes might even be good. Here's
an example. Several years ago I visited the pigs at two *multiplier
units* within the same company but located in different parts of the
country. A multiplier unit is a breeding colony that raises female pigs
to sell to farmers. The pigs all belong to the same genetic line, and
are very close to each other genetically because breeders use a fair
amount of inbreeding to keep consistency in the animals. The pigs
in these two units started out the same in every way, genetically,
physically, and emotionally.

But by the time I saw them, the pigs at the two units had developed completely different personalities. The pigs at one of the multi-
pliers were much more excitable and hyper, while the pigs at the
other one were calm and easygoing. It was the same story as the
rapist roosters: none of the humans had any idea they'd evolved a
whole new pig. They never visited each other's farms, so they didn't
know the pigs' personalities had diverged. That's what I mean by
natural—nobody was doing anything on purpose to affect the pigs'
evolution. It just happened.

When I took a look at the pigs' lives in the two units I found the
cause: the unit with the calm pigs had been selecting for placid temperaments unknowingly. In a multiplier unit each female pig has to
be individually evaluated for breeding stock. The employees weigh
her and look at her teeth, her udder, and the overall *conformation*
of her body. Conformation means that the different parts of the
animal are in good proportion to each other. The only difference
between those two units was that one unit had a good, stable scale
and the other unit had a crappy scale with a needle that jumped all
over the place. The unit with the crappy scale couldn't get a reading

on the more hyper pigs, and I'm sure they were culling the hyper ones and keeping the calm ones.

They didn't have a conscious *plan* to cull the hyper pigs; they just ended up doing it "naturally" because the scale was defective. Since the farm with the good scale could get reliable weights on all the pigs whether they stood still or not, there was no accidental selection pressure to get rid of the hyper ones. That shows you how sensitive animal genetics is to the environment: something as simple as how well the scale is working at a multiplier unit can change a pig's inherited emotional makeup.

Accidentally shaping a pig's genetics to create calmer animals is an example of a "natural" human selection pressure that's not only harmless but might be good for the animals. People and domestic animals have been together for a long time, and domestic animals have been evolving in response to humans for years. A pig wouldn't be a pig if it *hadn't* been evolving in the company of humans. He'd be some other kind of animal, like a wild boar. So probably a lot of the *incidental* selection pressures we put on animals are either harmless or good for the animal.

One genetic development I *am* concerned about, though: we're seeing more and more lameness in pigs today.

A Puppy Brain and Grown-up Teeth

Human selection pressure on animals' emotional makeup is probably the most obvious in dogs. I do not like what breeders are doing to purebred dogs. Breeders have made collie faces thinner and thinner, for example, leaving less and less space inside the skull for their brain. A dog needs a nice, wide skull to house its brain, and if you look at old paintings of collies from the beginning of the twentieth century that's what you see: Lassie with a broad, flat forehead.

By the early 1980s collie heads had gotten so narrow that a friend of mine who grew up with a collie on a farm in the 1950s and 1960s told me she didn't recognize the collies in her neighborhood as belonging to the same breed as her childhood pet. Somehow she got the idea that the "needle-nose collies" her neighbor owned came

from a whole different line of French collies she'd never seen before!
(I don't know how she got France in there, but she did.)

The problem isn't just the reduced space for the collie's brain; it's
also the weird shape of the skull. I would expect to find that the pro-
gressive narrowing of their faces has distorted collie brains anatomi-
cally. But whatever the cause, their intelligence has gone down so far
that I call collies brainless ice picks. It's a horrible thing to have
done to a nice, and beautiful, dog.

Making collies stupid wasn't the point, of course. Collie breeders
probably just wanted to exaggerate one of the most distinctive fea-
tures of the collie dog, which is its long thin nose. But in the process
of breeding for super-elongated noses, they bred out a decent-
shaped skull.

People probably put much more constructive selection pressures
on mutts. A mutt who bites people, or who destroys the house by
chewing everything in sight, has an excellent chance of being sent to
the pound or put to sleep. That means his genes will be removed
from the gene pool. Just about the *only* mixed-breed dogs who get to
reproduce are the ones who are well adapted to living with people—
and good at getting out of the yard. (I know owners are all supposed to neuter their pets, but a lot of owners don't. That's why we
have so many mutts.)

With purebred dogs the selection pressures are completely different, and a lot of them are negative. For one thing the breeders are
consciously trying to meet American Kennel Club standards, which
are heavily tilted toward physical criteria, not emotional or behavioral.
Moreover, professional breeders rarely if ever think about what their
beautiful dogs are going to do to an owner's house, and they usually
don't call the people who've bought their puppies to follow up on the
puppy's behavior. All the puppies in a litter could have emotional or
behavioral problems and the breeder could know nothing about it. So
there'd be nothing to stop him from continuing to breed the parent
dogs and producing more puppies with the same problems.

The other factor shaping purebred dog genetics is owners' higher
tolerance for difficult behavior in a beautiful, expensive animal. A
person who has spent $1,000 on a dog is going to put up with a lot
more bad behavior than a person who has spent nothing. And if the

owner is planning to breed the dog he won't stop to think maybe a dog who is horrible to live with shouldn't have puppies.

This is just a theory, but there's plenty of evidence on the emotional and behavioral problems of purebred dogs versus mixed breeds to support the hypothesis that the selection pressures on mutts are more constructive. For one thing, mutts are physically healthier, because the bad traits of purebreds, such as hip dysplasia, disappear just one or two generations away from the purebred line.

Mutts are also more likely to be emotionally stable, for a couple of reasons. One is that negative emotional traits will tend to get bred out of mixed-breed dogs, because mutts with major emotional problems like aggression or severe separation anxiety are probably more likely to be sent to shelters than purebreds. The other is that no one is practicing single-trait selective breeding with mutts, so mutts are not going to be turned into monster dogs the way the roosters were turned into monster rapist roosters.

In terms of behavior, the most important difference between mutts and purebred dogs is that purebreds are responsible for the large majority of fatal dog bites, not mutts. One twenty-year survey found that purebreds were responsible for around 74 percent of all fatal dog attacks on people. Seeing as how purebreds are only around 40 percent of the total pet dog population in the country, that's pretty bad.[4]

There have to be at least a couple of reasons for this. One, I'm sure, is that aggressive mixed-breed dogs get put down much more quickly than aggressive purebreds. But I think purebred dogs also suffer some of the negative emotional and behavioral effects of single-trait selection. Often breeders will mate their dogs so as to exaggerate one distinctive trait in the breed, like the rough collie's long thin nose. Since, as I mentioned earlier, any time you selectively breed for one trait, eventually you end up with neurological problems. Once you start getting neurological problems, one of those problems is likely to be aggression, so it doesn't surprise me that purebreds have more aggression problems than mutts.

Probably the most stable mutts are mixed-breed dogs whose underlying skin color isn't too light. Hair color doesn't matter; you just want to be sure you're not adopting an animal who has too many albino characteristics, such as blue eyes, a pink nose, and white fur covering most of its body. A small amount of white fur is perfectly fine, but you should avoid a white or light-colored coat combined with either blue eyes or a pink nose.

The fact that mixed breeds are so much less aggressive is good evidence that the selection pressures on mixed breeds are more constructive. My impression is that mixed-breed dogs are easier to live with in a lot of ways. I don't think there's any hard data on how much shoe chewing mixed breeds do compared to purebreds, but there's a lot of anecdotal evidence that purebreds, or at least certain breeds of purebreds, do more of it than mutts.

A friend of mine told me a typical story about mutts versus purebreds on chewing. Altogether she has owned two mixed-breed black dogs, and one yellow Labrador retriever. Labs, in case you didn't know it, are notorious chewers. Even though my friend got all three dogs as young puppies, the mutts did hardly any destructive chewing at all, whereas the yellow Lab chews everything she can get her teeth on. She's destroyed shoes, toys, pencils, pens, one corner of the rug in my friend's office, the fringe on the oriental carpet in the living room, three different wooden chair legs, several T-shirts that were left lying around, two blankets, a couple of books, several Tupperware containers, a sweatshirt, all the balls in the house, and she chewed the electric cord to the dehumidifier in half. That's just the list my friend could remember off the top of her head, and it's only the indoor list. Outdoors she wrecked a $400 hot tub cover, chewed up the wood framing around the neighbor's picture window, and, back in my friend's yard, gnawed clear through the trunk of a whole lilac tree, just like a beaver. The dog is one and a half years old now, and she's still going. She's doing a little better, because they've been working on training her to chew her rawhide bones instead of everything else, and because she's a little more mature. But she's still destructive. My friend estimates she's done $1,000 worth of damage at a minimum, not counting the two rugs that can't be repaired and would cost a lot to replace.

All Labradors chew like that; it's built into their genetic makeup. Golden retrievers are probably just as bad. I don't think anyone knows why, although with Labs it may be related to their obsessive overeating. (Although goldens belong to the same genetic group as Labs and all the other hunting dogs,[5] they don't overeat the way Labs do, so they may not chew for the same reasons.) I heard one owner call Labs "opportunistic eaters"—they'll eat just about anything you give them, including grapes and bananas. They love food so much you can train them to sit and heel just using little pieces of dry kibble. They are *always* hungry and will pack on the pounds if you let them eat whatever they want. I suspect it's this drive to eat that makes them need to chew everything in sight, because Holstein cows are the same way. Holsteins have been bred to eat a huge amount of feed so they can produce more milk, and they also have a compulsion to lick and mouth things and to manipulate objects with their mouths. Their behavior is so extreme that if a tractor gets left in their pen they'll lick the paint off and chew up all the hydraulic hoses. They'll destroy it, whereas beef cattle will just sniff it. Perhaps when you genetically increase an animal's desire to eat, you also increase its desire to use its mouth.

Of course that raises the question of why Labs have such massive appetites, and I don't know the answer to that. They were originally bred to be fishing dogs in Newfoundland, and they're impervious to cold and pain, so maybe the extra fat helps them stay warm. I don't know. That's why the quirks of selective breeding are so fascinating. If we knew why breeding a Lab to be a Lab also means breeding a Lab to be a compulsive overeater we might have a good idea of what causes most people to overeat, while some people can stop eating when they're full.

Getting back to mutts, I'd be surprised if any mutt ever chews things the way Labs do. Neither of the two black mutts my friend owned had her Labrador's chew-up-the-whole-house gene. Her first mixed-breed dog never chewed *anything* in my friend's house, even when he was a young puppy. The second dog went through a brief chewing period that he quickly outgrew. That dog was also highly responsive to punishment for chewing. He didn't have a strong inborn drive to chew things the way the Lab does, so he could easily stop chewing even with the haphazard "training" my friend gave him, which was mainly just to yell at the dog if she caught him chewing something he wasn't supposed to. That was all it took to get him to stop.

DOGS ARE PEOPLE'S
OTHER CHILDREN (NEOTENY)

The big problem with Labs is that they're permanent children. They're doing the chewing of a puppy but they're using grown-up teeth.

Humans have *neotenized* dogs: without realizing it, humans have bred dogs to stay immature for their entire lives. In the wild, baby wolves have floppy ears and blunt noses, and the grown-ups have upright ears and long noses. Adult dogs look more like wolf puppies than like wolf adults and act more like wolf puppies than wolf adults, too. That's because dogs *are* wolf puppies: *genetically, dogs are juvenile wolves.*

We know this thanks to Robert K. Wayne, a UCLA researcher who has studied mitochondrial DNA in wolves and dogs. There's only 0.2 percent difference between the mitochondrial DNA of a dog and the mitochondrial DNA of a gray wolf. The fact that dogs look so different from wolves doesn't mean anything at the genetic level; they're still wolves.[6]

Dr. Deborah Goodwin and her colleagues at the University of Southampton in the United Kingdom have done some very interesting research comparing dogs to wolves. She found that dogs who look most like wolves retain more wolf behaviors than dogs who have been bred to look as different from wolves as possible.[7] In other words, the more wolfie a dog looks, the more wolfie it acts. The King Charles spaniel, for instance, has lost half of the behavior patterns of wolves, and it still looks like a puppy when it becomes an adult.

I saw this up close in a dog I knew. He was a mixed-breed black-and-white dog with perfect, pointed ears and a long tapered nose, just like a wolf's. The strange thing about him was that he never, ever barked. He *could* bark, and he easily learned to "speak" (bark on command for food). But left to his own devices he didn't bark. He'd sit in the front bedroom monitoring the street, but when people came to the door he didn't launch into that crazed barking other dogs do. He'd get worked up and do a little "sneeze-bark," but that was it. I think that was probably his wolf ancestry showing through his dog exterior. Wolves don't bark, and neither did this wolfie-looking dog.

The King Charles study looked at the ages at which wolf puppies develop different aggressive behaviors, ranging from growling, which they can do by age twenty days, up through the *long stare*, which is the last aggressive behavior they develop, after they are thirty days old. You've probably seen pictures of wolves giving an enemy a long stare. They lock their eyes on to an animal's face and they stare fiercely. It's scary. I found a Web site written by a man who was on the receiving end of a long stare at an open zoo in England, the kind where you drive your car into a park where the animals live:

After a short while three wolves trotted up. They stood alongside the car window and just stared into my eyes. It was an unflinching, penetrating, calculated stare. It was a weapon designed to unsettle, not just an expression of interest.

The following day I had pretty much forgotten the lions and tigers, but I was still thinking about this long stare and I started to wonder why I had never seen a dog accomplish the same unnerving feat. After all, dogs are just domesticated wolves. So why does a Pekingese not have the ability to fix its owner with a withering stare?

Dr. Goodwin found that the reason a dog can't do a long stare is that dogs stop developing emotionally and behaviorally at the wolf puppy equivalent of thirty days. A grown-up German shepherd can do every aggressive behavior a thirty-day-old wolf can do, but nothing beyond that age. The only domestic dog Dr. Goodwin found who could do a long stare was the husky, which looks a lot like a wolf. A Chihuahua never advances past the wolf puppy equivalent of twenty days of age, so it's even more neotenized.

Dogs are the ultimate example of the accidental breeding programs humans create for the animals they live and work with. Many experts believe that one of the reasons wolves turned into dogs was that nursing human mothers probably adopted orphaned wolf cubs and nursed them at their breasts along with their human babies. Under this theory, the *only* reason dogs exist at all is that early people really loved wolf puppies, which gave any full-grown wolves who happened to have a case of arrested development a reproductive advantage. Humans got along best with submissive, puppylike wolves, and over time that's what they got, the same way the multiplier unit with the bad scale got calmer, nicer pigs.

The interesting question is whether dogs made us evolve into a different kind of human at the same time we were making them evolve into a different kind of wolf. I'll get to that later on.

ANIMALS AREN'T AMBIVALENT

Mammals and birds have the same core feelings people do. Researchers are just now discovering that lizards and snakes probably share most of these emotions with us, too. Just to give a couple of examples: the skink lizard in Australia is monogamous, and rattlesnake mamas here in the United States protect their young from predators the same way a mammal would. The fact that some snake mothers take care of their babies came as a big surprise, since researchers have always believed snakes weren't social at all and that mothers abandoned their babies after birth.[8] We still don't know much about the social lives of snakes, but at least now we know that they *have* a social life.

We know animals and humans share the same core feelings partly because we know quite a bit about how our core emotions are created by the brain, and there's no question animals share that biology with us. Their emotional biology is so close to ours that most of the research on the neurology of emotions—or *affective neuroscience*—is done with animals. When it comes to the basics of life, like getting eaten by a tiger or protecting the young, animals feel the same way we do.

The main difference between animal emotions and human emotions is that animals don't have *mixed emotions* the way normal people do. Animals aren't ambivalent; they don't have *love-hate* relationships with each other or with people. That's one of the reasons humans love animals so much; animals are loyal. If an animal loves you he loves you no matter what. He doesn't care what you look like or how much money you make.

This is another connection between autism and animals: autistic people have mostly simple emotions, too. That's why normal people describe us as *innocent*. An autistic person's feelings are direct and open, just like animal feelings. We don't hide our feelings, and we aren't ambivalent. I can't even imagine what it would be like to have feelings of love and hate for the same person.

Some people will probably think this is an insulting thing to say about autistic people, but one thing I appreciate about being autistic is that I don't have to deal with all the emotional craziness my students do. I had one fantastic student who flunked out of school because she broke up with her boyfriend. There's so much psychodrama in normal people's lives. Animals never have psychodrama.

Children don't, either. Emotionally, children are more like animals and autistic people, because children's frontal lobes are still growing and don't mature until sometime in early adulthood.[9] I mentioned earlier that the frontal lobes are one big association cortex, tying everything together, including emotions like love and hate that would probably be better off staying separate. That's another reason why a dog can be like a person's child: children's emotions are straightforward and loyal like a dog's. A seven-year-old boy or girl will race through the house to greet Dad when he comes from work the same way a dog will. I think animals, children, and autistic people have simpler emotions because their brains have less ability to make connections, so their emotions stay more separate and compartmentalized.

Of course, no one knows why an autistic *grown-up* has trouble making connections, since our frontal lobes are normal-sized. All we know right now is that researchers find "decreased connectivity among cortical regions and between the cortex and subcortex."[10] The way I visualize it is that a normal brain is like a big corporate office building with telephones, faxes, e-mail, messengers, people walking around and talking—a big corporation has zillions of ways for messages to get from one place to another. The autistic brain is like the same big corporate office building where the only way for anyone to talk to anyone else is by fax. There's no telephone, no e-mail, no messengers, and no people walking around talking to each other. Just faxes. So a lot less stuff is getting through as a consequence, and everything starts to break down. Some messages get through okay; other messages get distorted when the fax misprints or the paper jams; other messages don't get through at all.

The point is that even though autistic people have a normal-sized neocortex including normal-sized frontal lobes, our brains *function* as if our frontal lobes were either much smaller or not fully developed. Our brains function more like a child's brain or an animal's brain, but for different reasons.

When the different parts of the brain are relatively separate from each other and don't communicate well, you end up with simple, clear emotions due to compartmentalization. A child can be furious at his mom or dad one second, then completely forget about it the next, because being mad and being happy are separate states. A child hops from one to the other depending on the situation.

You see the exact same thing with animals. Strong emotions in animals are usually like a sudden thunderstorm. They blow in and then blow back out. Two dogs who live together in the same house can be snarling one second, then go back to being best friends the next. Normal people need a lot more time to get over an angry emotion, and even when a normal adult does get over a bad emotion he's made a lasting connection between the angry emotion and the person or situation that made him angry. When a normal person gets furiously angry with a person he loves, his brain hooks up *anger* and *love* and remembers it. Thanks to his highly developed frontal lobes, which connect everything up with everything else, his brain learns to have mixed emotions about that person or situation.

Another big difference between animals and people is that animals probably don't have the complex emotions people do, like shame, guilt, embarrassment, greed, or wanting bad things to happen to people who are more successful than you. There are different schools of thought about simple and complex emotions, but the definition I use is brain-based. Simple emotions are the primary emotions such as fear and rage that come from the reptilian and the mammalian brains. Complex emotions, or secondary emotions, also come from the reptilian and the mammalian brains, but they light up the neocortex as well. The secondary emotions build on the primary emotions and involve more thought and interpretation. For instance, shame, guilt, and embarrassment probably all come out of the same primary emotion of *separation distress*,[11] which I'll talk about shortly. Your culture and upbringing teach you when to feel shame versus when to feel embarrassment or guilt, but all three start out in the brain as the pain of being isolated.

I don't want to give the impression that animals *never* have more than one feeling at the same time. Later on I'll talk about the fact that cows often feel curious and afraid at the same time. (Jaak Panksepp, author of *Affective Neuroscience,* classifies curiosity as a core emotion.[12]) Biologically it's possible for more than one basic emotional system to be activated in an animal's brain at the same time, so technically an animal is capable of experiencing a mixed emotion.

But in real life one emotion usually ends up completely replacing the other, and some of the core emotions probably do "turn off" others. For instance, brain research shows that play and rage are incompatible emotions, which anyone who has ever watched two dogs play fighting can tell you. Once in a while a play fight will turn into a real fight, and when that happens the two dogs don't show the slightest sign (friendly tail wags, toothy smiles) that they're experiencing happy play feelings along with angry fight feelings.[13] Once a play fight has turned real, *all* of the dog's body language and vocal communication is angry.

NO FREUD FOR DOGS

Another huge difference between animals and people: I don't think animals have the defense mechanisms Sigmund Freud described in humans. Projection, displacement, repression, denial—I don't think we see these things in animals. Defense mechanisms defend against anxiety, and all defense mechanisms depend on repression in some way. Using repression, you push whatever it is you're afraid of down into your unconscious mind and focus your conscious mind on a stand-in. Or, in the case of the higher, more mature defense mechanisms, like humor, altruism, or intellectualization, you use humor, empathy, and thought to push away the "real" emotion, which is fear.

The reason I believe animals don't have Freudian defense mechanisms is that animals and autistic people don't seem to have re-pression. Or, if they do, they have it only to a weak degree. I don't think I have any of Freud's defense mechanisms, and I'm always amazed when normal people do. One of the things that blows my mind about normal human beings is denial. When I see a packing plant getting into a bad situation I'll say, "That's not going to work," and everyone will immediately think I'm being really nega-tive. But I'm not. It would be obvious to anyone outside the situ-ation that what they're doing isn't going to work, but people inside the bad situation can't see it because their defense mechanisms pro-tect them from seeing it until they're ready. That's denial, and I can't understand it at all. I can't even imagine what it's like.

That's because I don't have an unconscious. Normal people can push bad things out of their conscious minds into their unconscious minds, but I can't. Normal people can't always *keep* the bad stuff locked up, of course, but at least they have more freedom from it than I do. That's why I can't watch any violent movies with rape or torture scenes. The pictures stay in my conscious mind. Once they're there, I can't get rid of them. The only way I can block a bad image is by thinking about something else, but the bad image still pops back up in my mind, like a pop-up ad on the Internet. The way I think about it is that a normal brain has a built-in pop-up zapper, but my brain doesn't. To get rid of the pop-up image I have to con-sciously click on another screen.

I don't know *why* my brain doesn't have an unconscious, but I think it's connected to the fact that pictures are my "native lan-guage," not words. Lots of studies show that the language parts of your brain block your memory for images. Language doesn't *erase* your image memories; the images are still there, inside your head. But language keeps the images from becoming conscious. Psychol-ogists call this *verbal overshadowing*, and I'll talk about it more in my chapter on animal thinking. For the time being, let's just say that while I don't know why I don't seem to have an unconscious, I think my problems with language have a lot to do with it. Lan-guage isn't a natural ability for me, so maybe the language parts of my brain don't have the same power to overshadow the pictures.

I know it's a leap to go from saying that I don't have an unconscious to saying I don't have defense mechanisms, but based on my personal experience I think it's true. No one has ever tried to test animals for defense mechanisms, but animals act as if they don't have them, either. You never see an animal act as if a dangerous situation is safe. You might see a dog act like he's not afraid when he is, but that's not the same thing. The dog knows there's danger and is using a standard dog strategy to avoid provoking the threatening dog any further.

A friend of mine has two dogs, one a gentle female collie and the other a macho golden retriever. (You might not have thought a golden retriever could be macho, but this one is.) When my friend walks the collie *alone* past the two ferocious-acting German shepherds down the street, the collie looks straight ahead and acts as if she's deaf and blind. She does this because staring is a provocation. She's averting her eyes to avoid challenging them.

The reason we can say the collie is only pretending not to be afraid of the other dogs when she's alone, instead of not feeling fear because she's repressed it, is that she stops orienting to motion. All animals orient to movement. It's automatic. Since no dog can be oblivious to two German shepherds who are charging straight toward her, the collie has to consciously override her most basic orienting response. She has to *actively* ignore the other dogs.

FOUR CORE EMOTIONS

Researchers have identified and mapped out four primal emotions, all of which mature not long after an animal is born. They are:

1. Rage
2. Prey chase drive
3. Fear
4. Curiosity/interest/anticipation

Most animals also have four primary social emotions, which are not as well mapped:

1. Sexual attraction and lust
2. Separation distress (mother and baby)
3. Social attachment
4. Play and roughhousing

We know enough about fear, rage, and prey chase drive that these emotions deserve their own chapters: rage and prey chase drive in Chapter 4, fear in Chapter 5. For the rest of this chapter, I'm going to talk about curiosity/interest/anticipation and the social emotions.

CURIOSITY DOESN'T KILL CATS OR ANY OTHER ANIMAL

All mammals and birds are curious about and interested in their surroundings, and they *really* look forward to good things happening. You can see how much fun the state of anticipation is for an animal anytime you're getting a dog's food ready. All you have to do is start pouring dry kibble in a dog dish and your dog will break out in a huge doggy smile and begin wagging his tail at top speed. Getting-ready-to-eat is *always* a happy moment in a dog's life.

The dinnertime wag-and-smile come from one of the most basic emotions we have, an emotion that doesn't have a stand-alone name. You need at least two words to capture it, and even then you haven't completely got it. Dr. Panksepp says the best language he can come up with is *intense interest, engaged curiosity,* and *eager anticipation.*

When this brain circuit is activated, a person or animal probably feels some mixture of all three—curiosity, interest, and anticipation—depending on the situation. Humans have trouble describing what it's like to have this part of the brain electrically stimulated but usually "report a feeling that something very interesting and exciting is going on."[14] Animals who are having their curiosity-interest circuit stimulated act as if that's the way they feel, too. They get very animated and excited-acting and immediately start to run around like crazy, sniffing, exploring, and foraging.

Dr. Panksepp calls this part of the brain the SEEKING circuit.[15] Animals and humans share a powerful and primal urge to *seek* out what they need in life. We depend on this emotion to stay alive, because curiosity and active interest in the environment help animals and people find good things, like food, shelter, and a mate, and it helps us stay away from bad things, like predators.

We know that curiosity/interest/anticipation, or SEEKING, is a positive emotion from a field of research called *electrical stimulation of the brain*, or ESB. In ESB studies surgeons implant electrodes in an animal's brain and then watch the animal's behavior when different parts of the brain are stimulated. The SEEKING part of the brain is located mostly in the *hypothalamus*, which is in the mammalian brain, and the most important chemical involved is *dopamine*, which goes up when the hypothalamus is stimulated. The hypothalamus regulates sex hormones and appetite, so it makes sense that the SEEKING emotions would originate from there, since all animals spend a great deal of their time seeking food and mates.

We know animals like being in the SEEKING state because of *self-stimulation* studies where the researcher gives his animals control over the electrodes, so the animal can choose to turn the electrodes on or off himself. When electrodes are implanted into the curiosity/interest/anticipation system, animals turn them on and keep them on until they're totally exhausted from all their frenzied racing around and sniffing.

Since a lot of people read about these experiments in college I want to point out that the interpretation of these studies has changed completely in recent years. Researchers used to think that this circuit was the brain's *pleasure center*. Sometimes they called it the *reward center*. The main neurotransmitter associated with the SEEKING circuit is dopamine, so they thought dopamine was the "pleasure" chemical. That's what I was taught in college. When I learned about these experiments, I thought the ESB animals must be experiencing something like a permanent orgasm.

The pleasure center idea fit in with the fact that dopamine is involved in a lot of drug addictions. Cocaine, nicotine, and all the stimulants raise dopamine levels in the brain. Researchers assumed people develop addictions to drugs because drugs make you feel good, so dopamine must be the feel-good chemical in the brain.

But now researchers see things differently. We have a lot of evidence that the *reason* a drug like cocaine feels good is that it's intensely stimulating to the SEEKING system in the brain, not to any pleasure center. What the self-stimulating rats were stimulating was their curiosity/interest/anticipation circuits. *That's* what feels good: being excited about things and intensely interested in what's going on—being what people used to call "high on life"!

There are at least three different lines of evidence for this new interpretation. One is the fact that animals who are having this part of the brain stimulated *act* intensely curious. The second is the fact that human beings who are having this part of the brain stimulated *say* they feel excited and interested.

The third is the clincher. This part of the brain *starts* firing when the animal sees a sign that food might be nearby but *stops* firing when the animal sees the actual food itself. The SEEKING circuit fires during the *search* for food, not during the final locating or eating of the food. It's the search that feels so good.

That's not as surprising as it sounds when you think about it. At the most basic level, animals and humans are wired to enjoy hunting for food. That's why hunters like to hunt even if they're not going to eat what they kill: they like the hunting part in and of itself. Depending on their personalities and interests, humans enjoy any kind of hunt: they like hunting through flea markets for hidden finds; they like hunting for answers to medical problems on the Internet; they like hunting for the ultimate meaning of life in church or in a philosophy seminar. All of those activities come out of the same system in the brain.

ANIMALS LIKE NEW TOYS, TOO

In a natural setting, different animals have different levels of curiosity.

Rats, for instance, are super-explorers. They're very active and will explore every little nook and cranny of any environment you put them into.

Cattle are a lot less curious by nature, possibly because they're big enough, and have been domesticated for long enough, that they don't have quite so many dangers that need looking into.

Some cows are more curious than others. Holstein cows are very curious and do a lot of exploring with their tongues. If I lie down in the middle of a pasture filled with Holsteins they'll come up and start licking my boots. They'll go up to a horse, too, and lick him on his backside.

I wouldn't be surprised if we find out wild animals are more curious than domestic animals overall. My assistant, Mark Deesing, has a wild-type Carolina dog named Red Dog who is genetically closer to her wild ancestors than the AKC purebreds. Red Dog is *very* curious. If you take her to a new place she sniffs like crazy; she has to explore everything. One time we took her to a dog washing business that was next door to a McDonald's and she just went crazy sniffing and exploring. She wouldn't interact with us at all.

Mark's old dog, Annie, who was a little Australian blue heeler, was much more sociable. She was curious, too, but if we took her someplace like the dog washing business she would still interact with us. If domestic animals are less curious, it's probably the result of another difference in the selection pressures acting on domestic animals versus animals who fend for themselves in the wild. Domestic animals are taken care of by humans; they don't have to look for food or shelter. They don't have as much need for the emotion of curiosity as a wild animal does.

What I call *novelty seeking*—an animal's desire to touch and explore and interact with new things—is probably the same thing as curiosity. All animals like new things the same way people do. If you give an animal a bunch of nice toys to play with, and then a couple of weeks later you give him a brand-new, not-very-nice toy, he'll always prefer the new toy even if it's not as good as the old ones. That was true with my pigs at the University of Illinois. I gave them lots of good things to play with, like straw to root around in and telephone books for them to tear up. But if I brought in anything new they dropped everything and went for the new toy. All the pigs preferred a new, crummy toy, like a metal chain that you can't chew up, to an old, fun toy, like the straw and the telephone books.

That's why children always want new toys no matter how many toys they have already, and grown-ups always want new clothes and cars. It's the newness itself that's pleasurable.

So a liking for novelty probably comes out of the SEEKING system. It's curiosity for curiosity's sake, rather than curiosity for the sake of finding food or shelter. People and animals need to use their faculties, and curiosity is an important faculty. So people and animals need new things to stimulate their brains with. Parrots, for instance, need a huge amount of novelty to keep them from going stir-crazy. In one study of a single parrot, the more novel items the researchers put in the parrot's cage, the less likely the parrot was to develop feather-picking behavior, which is stress-related.[16] (Parrots also have to have lots of human companionship. They are highly social birds.)

For the time being, that's the best explanation I can come up with as to why novelty is both scary and fun. We need to know more about the SEEKING system in the brain. I do have one good story about this from a friend of mine, though. Her son is one of those children who hates change and transitions, so she's always working with him, trying to teach him to be more flexible. He loves new toys, but that's about it.

One day she was explaining to him that it was a little contradictory to be against new experiences while constantly bugging her to buy him new toys and he said, "I don't like new *things,* but I like new *stuff.*" That's exactly the way animals act.

ANIMAL SUPERSTITIONS

Curiosity doesn't just help animals find the things they need; it also helps them learn. That sounds obvious, I know, but the details of how curiosity helps an animal learn aren't.

It turns out that all animals and humans have what researchers call a built-in *confirmation bias.* Animals and humans are wired to believe that when two things happen closely together in time it's not an accident; instead the first event *caused* the second thing to happen.

For instance, if you put a pigeon in a cage with a key that lights up right before a piece of food appears, pretty soon the pigeon will start pecking the lighted key to get food.[17] He does that because his confirmation bias leads him to believe that the first event (the key lighting up) *causes* the second event (the food appearing). The pigeon happens to peck the lighted key a couple of times, the food appears (because food always appears when the key is lit), and *now* he concludes that *pecking* the key when it's lit causes food to appear.

The pigeon is acting like a person who thinks his team will win the baseball game if he's got his lucky rabbit's foot with him, which is why B. F. Skinner called this kind of behavior *animal superstition*. The pitcher has pitched a couple of good games while carrying his rabbit foot the same way the pigeon has gotten food a couple of times after pecking the lit-up key. In both cases, they've concluded that a correlation is a cause.

Confirmation bias is built in to animal and human brains, and it helps us learn. We learn because our default assumption is that if Event 1 is followed closely by Event 2, then Event 1 caused Event 2. Our default assumption *isn't* that Events 1 and 2 happened at the same time by coincidence. Coincidence is actually a fairly advanced concept both for animals and for people. That's why in statistics courses you have to formally teach students that a correlation isn't *automatically* a cause. Our brains are wired to see correlations as causes, period. Since in real life a lot of times Event 1 does cause Event 2, confirmation bias helps us make the connection.

The downside to having a built-in confirmation bias is that you also make a lot of unfounded causal connections. That's what a superstition is. Most superstitions probably start out as an accidental association between two things that aren't actually related to each other. You just so happened to wear your blue shirt the day you passed your math test; then maybe you just so happened to wear your blue shirt the day you won a prize at the fair; and after that you think your blue shirt is your lucky shirt.

Animals develop superstitions all the time thanks to confirmation bias. I've seen superstitious pigs. On farms, pigs get fed one at a time inside small electronically controlled feeding pens. Pigs can get into really nasty fights over food, so farmers use the pens to keep the peace. All the pigs wear electronic tags on their collars that work something like an electronic pass at a tollbooth. When a pig walks over to the feed pen a scanner reads the tag and opens the gate, then shuts the gate behind the pig so none of the other pigs can get in. The sides of the pen are solid, so the other pigs can't reach their snouts inside and bite the tail or rear end of the pig who's eating.

Once a pig is inside the pen she has to put her head up close to the feeding trough where another electronic scanner reads the ID, then measures out the exact amount of food that pig is supposed to eat.

Some of the pigs figure out that it's the collar that lets them into the feeding pen, and if they see a loose collar lying on the ground they'll pick it up and carry it over to the pen and use it to get inside. In that case confirmation bias has led them to the correct conclusion about the nature of reality.

But other pigs develop superstitions, also based in confirmation bias, about the feeding trough inside the pen. I saw several who would walk over to the feeder and go inside when the door opened, then approach the feed trough and start doing some purposeful behavior like repeatedly stomping their feet on the ground. They kept doing this until their heads happened to move close enough to the pen scanner to read their tags and deliver their food. Obviously they had had food delivered a couple of times when they happened to be stomping their feet, and they'd concluded it was the foot stomping that got them food. People and animals develop superstitions the exact same way. Our brains are wired to see connections and correlations, not coincidences and happenstance. Moreover, our brains are wired to believe that a correlation is also a cause. The same part of the brain that lets us learn what we need to know and find the things we need to stay alive is also the part of the brain that produces delusional thinking and conspiracy theories.

ANIMAL FRIENDS AND FAMILIES

On top of the four primal emotions, all animals and birds have four basic social emotions: sexual attraction and lust, separation distress, social attachment, and the happy emotions of play and roughhousing.

Sexual Attraction and Lust

Sex is another area where you see funky evolution thanks to human interventions. One example: American breeders have started selecting for much leaner pigs, because Americans want to eat leaner cuts of meat. So far the leaner pigs are healthy, but their personalities are completely different. They're super-nervous and high-strung. No one knows why this happens, although it might have to do with myelin, which is the fatty sheath surrounding the nerve cell axons that helps signals pass from one brain cell to another. Myelin is made of pure fat, so it's possible that when you breed a pig to have less fat you interfere with myelin production in some way. Lower myelin levels could produce jumpy animals because *inhibitory* signals—the chemical signals that tell other neurons *not* to fire—don't get through from one neuron to another. The animal can't calm itself down. That's one theory, anyway.

Lean pigs are also a lot less sexual. In China the pigs are all fat, and the mama pig makes way more piglets. A fat Chinese mother pig will have a litter of twenty-one piglets compared to just ten or twelve piglets in a lean American sow's litter. And the fat Chinese boars are super-sexy. When they brought them to the University of Illinois the boars would magically slip out of their pens and breed the sows whenever the staff wasn't around, something no American pig would do. They had nonstop sex on their minds and they turned into Houdini to have sex. All the fat Chinese pigs were super-calm and super-sexy. The females were really good mamas, too.

Sex is a very strong drive in any animal, so humans who take care of animals always have to be dealing with their sexuality one way or another. Either you want to prevent your animals from breeding, or you're trying to get them to breed successfully, and both goals have their challenges.

You can prevent unwanted breeding easily enough by neutering an animal, but you can't necessarily prevent all the *behaviors* that go along with breeding, especially not if you neuter an animal relatively late in its life after all its mature sexual behaviors have come in. That happened with our Siamese cat BeeLee when I was little. We neutered him pretty late, after his spraying behaviors were well established. One time we moved to a new house where we stacked all our pictures in the hall, waiting to be hung on the walls. BeeLee saw his reflection in the glass of the pictures, and he sprayed every single one. There were about thirty-five pictures altogether, and he completely ruined twenty of them. We had to throw them out. The rest were stinky, but we put them up anyway.

HOW TO MAKE A PIG FALL IN LOVE

Like all complex behavior, sexual attraction and mate selection depend on learning. The sex act itself is a hardwired *fixed action pattern*, like the rooster's courtship dance. It's hardwired into the brain, and an animal is born knowing how to do it. He doesn't have to be taught. But an animal does have to learn from other animals who he's supposed to mate with and who he's not supposed to mate with.

We know this partly because there've been so many stories over the years of animals who got mixed up in this area. There's a book called *The Parrot Who Owns Me*, written by an ornithologist at Rutgers University who adopted a thirty-year-old parrot after his owners died. The parrot got so attached to his new owner that he decided she was his mate. Every spring he would court her. He would shred newspaper to make a nest, he would kiss her, he would hoard food to share with her, and he would attack her husband if he saw him getting too affectionate with his wife. Then later on he'd act sorry for being mean to the husband.[18] There's also the famous story of *A Moose for Jessica*, about the moose in Vermont who fell in love with a Hereford cow named Jessica and courted her in her pasture for seventy-six days.[19]

Breeding domestic animals can be easy or hard, depending on the animal.

Cows and sheep are the easiest. Some cow and sheep breeding is done au naturel; they just send the males out in the pasture with the females, and they breed. The one thing you do have to be careful about with cattle is dominance hierarchies with the bulls. The most dominant bull doesn't necessarily have the best semen or the best genes. So if the top bull is shooting blanks and chasing off all your good bulls, that's bad. You have to try to put enough bulls out with the cows that one dominant bull won't breed them all.

Most of the dairy cattle breeding is done by artificial insemination, which is easy with cattle. You don't have to do anything special with the females. You just thread a catheter into their wombs and inject the semen and that's it.

There's a little more involved with some of the bulls, especially the Brahman bulls. Those are the white cattle with the big humps on their backs and the long ears. Brahman cattle are very affectionate toward people, and they love to be petted. They just eat it up. I love Brahman cattle. If you treat them nice, they'll treat you nice. They'll lick you all over your face and body. But if you treat them bad, look out. They'll kick you or charge you.

Brahman bulls are so affectionate that when you collect semen from a Brahman bull you have to pet them a *long* time first. They'll refuse to give the semen for twenty minutes because they want twenty minutes of throat and butt scratching; that's the stuff they really care about. Then they'll give it to you. They'll delay the sex in order to get some good, serious stroking. With some of them you have to walk away and leave or they won't give you the semen at all. You have to let them know that if they don't give the semen they're not going to get stroked.

Pigs could be bred naturally, too, but a lot of the time breeders use artificial insemination instead. Breeding pigs commercially is an art. I talked to a man who had one of the most successful records for breeding sows out there and he told me things no one's ever written in a book as far as I know. Each boar had his own little perversion the man had to do to get the boar turned on so he could collect the semen. Some of them were just things like the boar wanted to have his dandruff scratched while they were collecting him. (Pigs have big flaky dandruff all over their backs.) The other things the man had to do were a lot more intimate. He might have to hold the boar's penis in exactly the right way that the boar liked, and he had to masturbate some of them in exactly the right way. There was one boar, he told me, who wanted to have his butt hole played with. "I have to stick my finger in his butt, he just really loves that," he told me. Then he got all red in the face. I'm not going to tell you his name, because I know he'd be embarrassed. But he's one of the best in the business—and remember, this *is* a business we're talking about. The number of sows successfully bred by the boars translates directly into the profits a company can make.

This same man also told me he had to deal with the female pigs the same way. With a cow you can just take a catheter and insert it into her womb and she'll have babies. She doesn't have to be turned on or interested. But you have to get the sow turned on when you breed her so her uterus will pull the semen in. If she isn't fully aroused she'll have a smaller litter because fewer eggs will get fertilized.

So the breeder has to be able to tell exactly when the female pig is ready. One of the signs you look for is that when a pig is sexually receptive her ears will go "blink!" and pop straight up. That's called *popping*. Also, when you put pressure on her back, which is what she would feel when the boar mounts her, she'll stand perfectly still. Breeders call that "stand for the man." A good breeder knows when his sows are ready to stand for the man, and he usually sits on each sow's back when he inserts the semen so she feels that pressure on her back. Some breeders put weights on the sow's back to accomplish the same thing.

Pig breeders used to ignore all these psychological factors, but now they pay attention. One thing that's really important: the man who does the breeding *cannot* be involved with any nasty things, like vaccinations or any kind of veterinary care. (Nasty from the pig's point of view, I mean.) If he does any of that stuff, the pigs will reject him. He might still be able to breed them, but they'll have smaller litters. Paul Hemsworth, from Australia, showed that sows who are afraid of people have 6 percent fewer piglets than sows who aren't afraid of people, and the piglets don't do as well on weight gain after they're born.[20] The people attending the farrowing also have to be people the pig trusts completely. So the employee handling the breeding has to do only the breeding and nothing else.

HORSES IN SUPER-MAX PRISONS

Pig breeders respect the animals' nature, and they do a good job with their animals. But I have a lot of complaints about horse breeders. They keep the stallions locked up alone in their stalls all day long, where they go crazy with nothing to do and no one to interact with. Horses are social herd animals, and they need to be with other horses. The super-max prisons we keep stallions in distort their sexuality.

Out on the range, a stallion who wants to mate a mare walks up to her and whinnies. He's saying, "Would you like to have sex?" and he has to ask very nicely to breed her. If the female doesn't cooperate he isn't going to get anywhere.

But a stallion who's been locked up in a stall turns into an aggressive sex maniac. The mating procedures owners use are horrible. They tie up the mare so she can't run away, and then they hobble her feet so she can't kick the stallion if she doesn't like him. Then they let the stallion out and he just runs up to her and rapes her. It is disgusting.

I understand why the breeders don't want to do things the natural way. They're afraid the mare will kick the stallion and injure him. But turning stallions into horse rapists is wrong. It's completely abnormal, and keeping the stallions locked up the way they do is terrible. A racehorse who's been reared in isolation probably does need his own stall for protection, but that's because his character has already been warped. Horses don't need private stalls; they need other horses. The owners may be sparing no expense providing food and shelter, but they're just not thinking.

HORMONES OF LOVE

We know a fair amount about the brain basis of sexuality. Everyone has heard of testosterone, estrogen, and progesterone, and probably most people know that both sexes have all three hormones, though in different amounts. Two other important hormones aren't as well known: *oxytocin* in females and *arginine vasopressin* (AVP), or *vasopressin,* in males. (Some readers may have heard of AVP from their pediatricians. AVP is also called antidiuretic hormone, or ADH, because it increases water retention. Doctors sometimes prescribe it for children who wet the bed.)

Oxytocin shoots up right before a mother animal gives birth and helps her be a good mother, and both oxytocin and vasopressin shoot up in male brains as well as female brains during sex. (Oxytocin is more important in the female brain and vasopressin is more important in the male brain.) These are very, very old chemicals. Both of them evolved from *vasotocin,* which controls sexual behavior in frogs and other amphibians. If you put just a little bit of vasotocin into a frog's brain the frog will immediately start performing courtship and mating behaviors. There's only one amino acid difference between vasotocin, oxytocin, and vasopressin, so when it comes to sex, we still have our frog brains working for us.

Vasopressin and oxytocin aren't just sex hormones. They're motherhood, fatherhood, and love hormones, too. Some science writers have called vasopressin the monogamy hormone, because prairie voles, who mate for life, have much higher levels of vasopressin than their cousins the montane voles, who don't mate for life. (Only 3 percent of all mammals are monogamous.) Mother and father prairie voles build nests together and raise their babies together. Thomas Insel, a neuroscientist who has done a lot of the research on vasopressin and voles, has found that when you put high-vasopressin prairie voles together in a big roomy cage the male and female mates will spend half their time close together. When you put low-vasopressin montane voles inside the same cage, they spend almost all of their time alone and only 5 percent of their time physically close to another vole.[21]

Oxytocin is especially important to all these social activities, because oxytocin is essential to *social memory*: oxytocin is the hormone that lets animals remember each other. An experiment with mutant mice who didn't have the gene for oxytocin found that the mice didn't form social memories. They could remember everything else just fine, but they couldn't remember that they'd already met a mouse who had just been put in their cage, and they'd start sniffing him like he was a total stranger.[22] (Animals who already know each other never sniff each other as much after a separation as they do when they first meet. You can see this easily with dogs.) Obviously, if you don't have social memory you can't be monogamous, and you're not going to be a very devoted mom, either, if you have trouble recognizing your babies.

This finding has led researchers to speculate that some autistic people might have faulty oxytocin production, since a lot of times autistic people don't seem to remember people they've met before, either. However, a lot of that has to do with *face recognition*, which is extremely poor in autistic people, not face *memory*. This is another aspect of autism that no one understands, although we do have brain scan data confirming it. There's also a study showing that normal people use different parts of the brain to recognize an object versus a face, whereas autistic people use the *object recognition* area of the brain to recognize objects *and* faces. I have a terrible time recognizing people's faces myself, but I don't have any trouble remembering people in other ways, like through their voices. Oxytocin might be involved in autism; I don't know. But I'm guessing that autistic people's face recognition problems come from something else.

Vasopressin also makes prairie voles sexually possessive. They *mate guard*, which means they stick close to their mate and fight off any other male who approaches. They're more territorial, and they're much more aggressive toward other males even when their mates are not present. One study looked at the relationship between vasopressin and *intermale aggression*, which is a male animal's tendency to attack another strange male that is put into a cage with him.[23] The researchers found that adult male voles who are still virgins are almost never aggressive. But once they've mated and had their vasopressin levels rise they "exhibit a long-lasting, permanent increase in aggression." In the study the researchers injected newborn prairie voles with vasopressin over the first seven days of life, then tested them for aggression. The treated voles were much more aggressive, not just toward other males but even toward females.

The montane voles, who have low vasopressin, could care less about their mates or about other males. Once they've mated a female they disappear. They're just not very socially motivated. The montane females are loners, too. Oxytocin is the *maternal hormone*, and montane females have lower levels of it than the prairie females. Montane females abandon their babies soon after they give birth— and the babies aren't too bothered by this, because they aren't very social, either.

Compare that to the way a mama dog acts with her babies. One time Mark's dog Annie accidentally got locked in the kitchen and she couldn't get to her puppies, who were in the attached garage. Annie went crazy. First she violently scratched the door, then she attacked the plasterboard on the wall between the kitchen and the garage. She was so frantic that she clawed clear through the wall into the garage. Annie was a relatively small dog, only thirty to thirty-five pounds, but she was so desperate to get to her babies she tore through a wall.

Dogs probably have fairly high oxytocin levels. They're highly social animals to begin with, and an animal has to have good oxytocin levels to be highly social. Wolves are often monogamous, and even when they're not strictly monogamous they practice serial monogamy. The dingo and the Carolina dog are usually monogamous, too.

On the other hand, domestic dogs don't look like they're monogamous at all. A male dog on the loose will mate any receptive female he finds and then go tearing off to find any other receptive females in the area. However, that might be due to the fact that dogs never become full adults emotionally, so they don't develop an adult wolf's capacity for monogamy. Also, we don't really know what domestic dogs' social life would be like if they didn't live with people. Very few pet dogs have the option of mating with another dog for life.

A dog's oxytocin levels rise when his owner pets him, and petting his dog raises the owner's oxytocin, too. I'm sure that's one reason why so many people have dogs in the first place. I don't think anyone has researched this yet, but I expect we'll find that dogs make humans into nicer people and better parents. Oxytocin is definitely important in humans. When women have babies their oxytocin levels shoot up right before the birth, and research shows that those high levels spark maternal warmth and care. Oxytocin produces caring "maternal" behavior in men, too. So for parents, owning and petting a dog is probably like getting a "good parent" shot every day. Dogs are probably good for marriages for the same reason.

One of the interesting things about the research on vasopressin is the way behaviors we humans tend to think are "bad," like aggression and sexual possessiveness, go together with behaviors we think are good, like taking care of the young and being faithful to your mate. Male prairie voles have higher aggression and higher mate guarding, and they're also faithful husbands and nice dads. Male montane voles don't have much aggression or any mate guarding at all, but they're promiscuous and totally uninterested in their offspring. Take away the aggression and the mate guarding and you lose the devoted mate and the good dad, too. They go together.

The research on testosterone and paternal behavior isn't as clear as the research on vasopressin. A lot of researchers have concluded that testosterone lowers paternal behavior, but the most recent research shows that in a *monogamous* animal, testosterone increases paternal behavior. The body converts testosterone to estrogen, and the estrogen increases nurturing of the young.

ANIMAL LOVE

All baby animals make a high-pitched distress call when they're separated from their mothers. (I don't know whether montane vole babies have a distress call, but I assume they probably do if only for a short period of time.) Animal babies are totally attached to their mamas, and when they grow up most animals are strongly attached either to a particular friend or to the members of their social group, or both. Animals love other animals.

Animals make social distinctions between friend and stranger the same way people do, too. I heard a story about a guy who was stealing pigs at an auction a while back. Farmers bring their hogs to auction to sell to buyers from the packing plants. The auctions last for a few days and handle thousands of pigs, so it would be easy to take just one or two pigs a day without anyone noticing, which is what the thief was doing. The only reason they knew someone was stealing was that the trucks were coming up short. A truck holds two hundred animals, and when a farmer delivered a truckload of pigs to the loading dock the stockyard manager would do a head count and find that one pig was missing.

They discovered who the thief was when somebody noticed a pen where none of the pigs were lying together. Each pig was keeping his distance from the others, and the guy who noticed them realized that the pigs in that pen were acting like strangers. The reason they were acting like strangers was that they *were* strangers. They'd come from different farms.

The thief turned out to be an employee who was taking one or two pigs a day out of the thousands at the auction and moving them to a pen in the back where he was keeping them until he could take them home. The pen looked like every other pen, and there would have been no reason for anyone to think the pigs inside had been stolen if the pigs themselves hadn't known they didn't belong there. The pigs' behavior gave him away. They weren't with their friends, and they acted like they weren't with their friends.

People constantly underestimate domestic animals' need for companionship. A good way to understand just how social these animals are is to ask yourself how horses, cows, pigs, sheep, dogs, and, to a lesser degree, cats, came to be domesticated in the first place. Why did wild horses decide it was okay to have people sitting in a saddle on their backs holding a pair of reins? It's pretty incredible.

Most experts believe that the reason these animals became domesticated was that they were highly social. Their innate sociability led them to associate with humans and eventually to accept human ownership and direction. That's a high degree of sociability, and it's still there in all of our domestic animals. Even cats are much more social than people realize; sister cats even help each other give birth. *All domestic animals need companionship*. It is as much a core requirement as food and water.

Some ranchers are beginning to take this into account. In the past I've watched calves being separated from their mamas here at the university when they reached weaning age, which is three to six months. There's a lot of individual variability in how the calves and the mothers react, and some of them would get horribly upset. I remember one mama who was mooing frantically and trying to jump the fence to get back to her baby. The babies acted really stressed and agitated, too.

Now people are starting to do low-stress weaning, where the mothers and the babies are separated by a fence but they can still touch noses. That's all the babies care about by that age. They don't care about the nursing; they care about being together with their mom. If you didn't separate the calves, and just let nature take its course, female calves would probably stay with their mothers for good. You see that a lot in the wild, mothers and daughters staying together. You also see males stay with their brothers, and in some species males form friendships with other males.

A dog's attachment to his owner is like a baby animal's attachment to his mother, or a human child's attachment to his mom or dad. Pet dogs act the exact same way children do in the *strange situation test*. In the strange situation test the researcher watches how a very young child reacts to a strange new environment when his mother is there with him, and when she's not. Most children will confidently explore a strange environment as long as their mother is with them, but when she leaves the room they'll stop exploring and wait anxiously for her to come back. Dogs do exactly the same thing. This has been tested formally in fifty-one dogs and owners. Most dogs stop exploring and act anxious when their owner leaves the room. Then they relax and start exploring again when their owner returns. When humans say dogs are like children, they're right.

Researchers do ESB—electrical stimulation of the brain—research on social attachment by recording which areas in the brain cause an animal to make separation distress calls when stimulated by electrodes. Using this technique they've been able to map out these circuits and the chemicals that are involved pretty well. Evolutionarily, social distress is linked to three old, primitive systems in the brain:

1. *Pain response.*
2. *Place attachment:* the animal's ability to form an attachment to its nest, breeding territory, or home. (Babies of all species are distressed when left alone, but they're less distressed if they're left alone in their home, not in a strange place.)
3. *Thermoregulation:* the regulation of body heat.

You can see all three in the language people use to talk about their social attachments, and in the way they act.

The connection between social separation and the pain system is probably the most obvious in our language. We use the same words to describe physical pain and social separation and loss: *pain, anguish, agony,* even *torture.*

Place attachment you can see in sayings like, "There's no place like home."

Thermoregulation comes up all the time when people talk about relationships. We use the expression "maternal warmth," and we say people are warm or cold. Warm people are loving, kind, and connected, and cold people are the opposite. Also, people and animals who are feeling lonely usually want to be touched, which comes from the fact that in the wild babies keep warm by staying close to their parents' bodies.

I know that sounds strange, but researchers believe that social warmth evolved out of the brain system that handles physical warmth. That should tell you something about how important social attachment is to animals. In all mammals a baby has to have a strong social attachment to its parents in order to survive. A baby wolf needs social contact to stay emotionally warm as much as it needs physical contact to stay physically warm. *Social attachment is a survival mechanism* that evolved partly from the survival mechanism of keeping the body warm.

LOVE HURTS

The same chemicals in the brain are involved, too. Most people know that the brain has its own painkillers called *endorphins*. Endorphins are *endogenous opioids*: they are nature's version of morphine and heroin. The brain circuit that releases endorphins is called the *opioid system*. The brain releases endorphins when we're injured to reduce pain, and also when we are with people we love, or when someone we love touches us. A lot of neuroscientists think we probably become addicted to or dependent on people in a similar way to heroin or morphine addiction, too. People who are attached to each other develop a social dependence on each other that's based in a physical dependence on brain opiates.

A lot of interesting research has been done on the *opioid theory* of love and friendship using the drug *naltrexone,* an *opioid antagonist* that blocks opiates in the brain. Jaak Panksepp is best known for this research, and I've also done an experiment with Nicholas Dodman, who is a professor in the Tufts University School of Veterinary Medicine and also the author of *The Dog Who Loved Too Much*. Doctors use naltrexone to treat heroin and alcohol addictions, but it also blocks endorphins, so researchers can use naltrexone experimentally to see what happens to social attachment and distress calls when an animal's opioid system isn't working.

What researchers have found is that animals get much more social when they take naltrexone, which is exactly what they hoped they'd find.

Here's how it works. Naltrexone blocks the effects of opioids in the brain, which feels bad. Functionally speaking, having your opioids blocked is the same as having low opioids. Social contact raises opioids in the brain, which feels good. In theory, animals who've taken naltrexone ought to get more and more social, because they are trying to raise their endorphin levels back up to where they were before the naltrexone blocked them. An animal whose endorphins are low should try to raise his endorphins by getting more social contact the same way a heroin addict whose heroin levels have gotten low will want to use more heroin.

That's what happened in the experiments. Dogs on naltrexone wagged their tails more and monkeys groomed each other more. Animals taking naltrexone become more sociable.

I haven't heard of any experiments on naltrexone in typical humans, but Dr. Panksepp has done a lot of work using naltrexone in autism, because he thinks some autistic people may have *too many* natural opioids in their brains. *High* levels of opioids lower social desire, which is why heroin and morphine addicts withdraw from social contact. They stop feeling the need for other people. Dr. Panksepp thinks some autistic children may be like heroin addicts. They don't feel the need to interact with other people because their opioids are too high.

He based this theory on the fact that some autistic children have abnormally low pain sensitivity, which might come from too-high opioid levels, and also on the fact that some autistic children do not cry real tears. When you give opiates to animals they don't cry at all, so it's possible an autistic child who cries without tears might have a problem in his opioid system.[24] (Animal researchers use the same word to describe crying in animals as in people.) Dr. Panksepp also thinks autistic children who like very spicy, salty, or hot foods might have naturally high opioid levels, so they might get more social tak-

ing naltrexone. So far he's found that about half of the autistic children he's treated have gotten more social when they take low doses
of naltrexone.[25]

FEELINGS FROM MY SQUEEZE MACHINE

When I saw the cattle in their squeeze chute and got inspired to
build a squeeze machine for myself, at first I was thinking only about
the calming effects of deep pressure. So I built it with just two hard
plywood boards, without any padding or cushions. All autistic children and adults like deep pressure. Some of them will put on really
tight belts and hats to feel the pressure, and lots of autistic children
like to lie underneath sofa cushions and even have a person sit on
top of the cushions. I used to like to go under the sofa cushions
when I was little. The pressure relaxed me.

Then gradually I started to improve my squeeze machine by
adding soft padding to the boards, and I got a second feeling that
was different from just feeling relaxed and calm. The pads gave me
feelings of kindness and gentleness toward other people—social feelings. It also made my dreams nicer. I would have dreams about petting puppies, or being out in the green pasture at my aunt's ranch
with the blue skies overhead. Things like that. The hard boards
made me feel physically calm, but the soft padding made me feel
social. I had to have the nice feeling of being held to have nice
thoughts about people.

My experience reminds me of the famous experiments back in the
1960s by Harry Harlow at the University of Wisconsin. He tested
baby monkeys to see whether they would prefer fake mothers made
of wire or of soft cloth. All the babies preferred the soft cloth
mother, even when all the milk came from the wire mother. The
contact comfort was more important to them than the food.

I noticed something else with my padded squeeze machine, too.
After I started using it, I would sometimes feel worse the next day,
more anxious. That didn't happen with the hard plywood. Today I
think I probably felt more anxious because the soft squeeze machine
was activating my opioid system. It was making me have nice social
feelings, and it was also making me physiologically dependent on my

machine the same way people are dependent on social contact to keep their endorphin levels up. When I used the squeeze machine and raised my opioids, then didn't use it again for a few days, I was having a withdrawal. I was developing a social dependence on my squeeze machine.

I think the squeeze machine probably also helped me have more empathy, or at least more empathy for animals. When I first started using the soft version of the machine, in my late teens, I didn't know how to pet our cats so they really liked it. I always wanted to squeeze them too tight. Then after I used the soft machine I thought, "I have to make the same feeling I have go to the cat." I walked out of the room and the cat was in the hall, and I started stroking him, trying to transfer the feeling I had in the machine. Before I used the soft squeeze machine BeeLee used to run away because I always squashed him. But that day he started purring and rubbing up against me. I realized, "I know how to pet kitties so they like me." This happened immediately after I used the soft machine for the first time. I remember the exact moment I did it.

Autistic children never know how to pet animals the right way, so you have to teach them. Usually they want to squeeze the animal way too tight. I talked to one young woman with Asperger's syndrome about her pet cat. Asperger's syndrome is a form of autism where the person's IQ is normal and language comes in on time. (A person with autism can also have a normal or a high IQ, but to have the diagnosis of autism you have to have had a language delay.) She told me her cat didn't like to be squeezed, but since she liked to squeeze the cat she kept on doing it. I told her, "You must not squeeze the cat," and I stroked her arm to show her how she had to touch her cat.

Even a lot of normal people don't realize that you have to *stroke* animals, not pet them. They don't like to be petted. You have to stroke them the way the mother's tongue licks them.

So far no one has studied empathy and the opioid system. Researchers have just measured things like distress calls. But my experience with the cat might mean that *social intelligence* may be partly based on the opioid system, too, not just social attachment and dependence.

SQUEEZE MACHINES FOR
PIGLETS AND BABY CHICKS

There have been two experiments on squeeze machines for animals, one by Dr. Panksepp and one that I did with Nick Dodman. In Jaak's experiment, he hollowed out a little foam square to make a squeeze machine for a one-day-old baby chick. He put a little fluffy chick inside the foam square with its head sticking out, and then counted how many distress calls the chick made when it was separated from its mother. Inside its foam square the chick cried a lot less, which is what Jaak predicted would happen.[26] That's evidence that the soft squeezing raised the chick's brain opioids, since animals stop crying when they are given opiates. It's not just social contact that raises endorphins in the brain. *Social touching also raises endorphins.*

The study I did with Nick didn't turn out as neatly. We built a squeeze machine for a piglet out of two boards wrapped in foam rubber and covered by gray plastic upholstery material. The two boards were inside a little pen with a gate at the front, and on the other side we had another piglet standing there facing the piglet inside the squeeze machine so they could touch noses through the gate. We had to have the other piglet because if we didn't the lone piglet would just go crazy with anxiety.

We didn't actually squeeze the piglet in the squeeze machine, because we wanted to see whether he would *squeeze himself* against the foam boards after we'd given him naltrexone to block his opioid system. We were predicting that he wouldn't squeeze himself against the foam, because he wouldn't be able to feel any of the good endorphins that come from physical contact. Squeezing himself wouldn't feel any better than not squeezing himself.

Normal baby pigs love to snuggle into each other. If they get anxious or excited during handling they stick together so tightly you can't get them apart. Hog farmers call it "squealing super-glue." So if you put a normal piglet inside a piglet squeeze machine he'll snuggle down against the foam and go to sleep, probably because his endorphins help put him to sleep. It's a little like a heroin addict "nodding out," only it's natural and healthy. (Dr. Panksepp has done an experiment with baby chicks showing what happens if you raise their opioids by giving them a low dose of an opiate drug and then hold them in your hand. The baby chick will stop cheeping, snuggle right down, and close its eyes. Raised endorphins probably have exactly the same effect.)

So we predicted that naltrexone would prevent the piglet from snuggling into the foam and going to sleep, because he wouldn't be able to feel the effects of higher endorphins in his brain, and he wouldn't get sleepy. He would stay awake and stay standing up.

That wasn't exactly what happened, though. At first the piglet couldn't settle down at all, so we were right about that. But after a while he managed to do it. All the naltrexone seemed to do was delay the contact comfort response. Jaak Panksepp found the same thing with his baby chicks. Even when he completely blocked their opioid systems, they still settled down eventually.

I think the explanation may have to do with oxytocin. Oxytocin also goes up with physical contact, and I think what might have happened is that every time the pig briefly squeezed himself against the soft foam he probably raised his oxytocin a little more until finally his oxytocin levels were high enough to compensate for the missing opioids and he settled down.

This research is important for people with autism. A lot of autistic children can't stand to be touched. I was like that when I was a little kid. I wanted to feel the nice social feeling of being held, but it was just too overwhelming. It was like a tidal wave of sensation drowning me. I know that doesn't make sense to people who aren't autistic, and the only other way I can think of to describe it is being in the ocean with waves washing over you that keep getting bigger and bigger. At first the waves feel good, and the sensation is soothing and relaxing. But as the waves get stronger and more powerful you feel like you're starting to drown and you panic.

Being touched by another person was so intense it was intolerable. I would start to panic and I had to pull away.

That's why I would get under the sofa cushions, because I could control those. I could let that good feeling wash over me, and if it got too intense I could stop. But when people hugged me they wouldn't stop. I had this one very affectionate big, fat aunt and when she hugged me she had this horrible perfume so there was that smell overwhelming me along with the touch. I had to get away.

When I first used my squeeze machine it was overwhelming, too. I had to force myself to relax into it and let the good feeling wash over me. Today I think it's very important to desensitize autistic children to touch, because all children need to be touched. It's not that autistic children don't want to be touched; it's that their nervous systems can't handle it. A lot of occupational therapists have ways to work with an autistic child so that touch starts to feel much less intense and more normal. That's important.

Nonautistic people can have problems with touch, too, of course. Over the years, what's interesting is that I've found that some guys hate the idea of the squeeze machine because they don't want to give in to it. Girls always like it better than the guys. Big macho guys especially don't like it, and anyone who's claustrophobic hates it. With men, I've found that a lot of guys also don't know how to pet animals right. They'll pet them too roughly, and the animals don't like it. A lot of times men play too roughly with dogs, too, at least in my experience. I don't know whether petting is related to opioids or oxytocin or maybe to both, but men have lower levels of oxytocin than women, so maybe when a man has a rougher way of touching animals than most women, it's related to oxytocin. I don't think it's clear whether men have overall lower levels of oxytocin, but their testosterone may make them less responsive to the oxytocin they do have.

ANIMALS LOVE TO PLAY

Nobody knows why animals love to play so much, but they all do. That's where the emotion of joy comes from, the play circuits in the brain. When big old huge dairy cows are let out in the spring, after spending the whole winter cooped up in the barn, man, they just jump around all over the fields like little calves. It's the same feeling young animals have when they play.

We don't know as much about the brain basis of play as we do for curiosity, love, and sex. One thing we do know is that you don't need any neocortex at all to play. That's not to say the neocortex never lights up during play; it probably does. But if you remove the neocortex, an animal will still play. And if you damage the frontal lobes, which are the decision-making, responsible part of the neo-cortex, an animal actually plays *more*.

That fits in with the fact that all human children do less and less roughhouse play as they grow up and their frontal lobes mature. Probably the more dominant your frontal lobes, the more "serious" and nonplayful you are. Stimulants increase frontal lobe functioning, so naturally they would decrease play, too. As a matter of fact, some parents whose hyperactive children (ADHD, for attention deficit hyperactivity disorder) take Ritalin or other stimulants complain that their kids lose too much of their playfulness, and when you give a stimulant to a young animal it plays less, too. So play is definitely not a neocortical function.

There are other chemicals that decrease play, including the stress hormones and oxytocin. We also know that lots of opioids are released during play. But none of this adds up to a clear picture of the brain biology of play, and behavior studies don't tell us why animals play, either. But the fact that they all play at just about the same age relative to brain development tells us that play may be important to brain growth and/or to socialization.

Two researchers, John Byers at the University of Idaho and Curt Walker at Dixie State College of Utah, have developed an interesting theory about *locomotor play.* Locomotor play is the pretend chasing and jumping-and-spinning play a young animal does when it's alone. (If you want to see locomotor play, watch a goat. They are the biggest jumpers and spinners ever.) Drs. Byers and Walker think the purpose of locomotor play might be to help grow good connections among the cells in the *cerebellum,*[27] which is the small, round "ball" down at the bottom of the brain that handles posture, balance, and coordination.[28] Their research shows that in mice, rats, and cats, locomotor play begins when the cerebellum starts to form lots of new connections (or *synapses*) between its cells, and peaks at the same time synaptic growth peaks. So mice start locomotor play at around fifteen days after birth and hit their play peak from four to ten days later; cats begin locomotor play around four weeks after birth, and reach their play peaks at around twelve weeks. In both mice and cats, the peak point of brain growth is also the peak point of locomotor play.

Since the cerebellum handles physical coordination, it makes sense that a young animal or human might spend a lot of time leaping, running, and chasing during the period that his cerebellum is forming new connections. The locomotor play period also coincides with the period when muscle fibers are turning into either *fast-twitch* or *slow-twitch* fibers. (Fast-twitch fibers give you the kind of short-lasting blasts of power you need for sprinting; slow-twitch fibers give you the long-lasting, endurance strength you need to run a marathon. Your heart would have to have slow-twitch fibers, or you'd be dead.)

So far this finding is only a correlation, and we don't know from a correlation whether it's the locomotor play that's causing the cerebellar development, or the cerebellar development that's causing the locomotor play, or both, or neither. Researchers will have to run controlled experiments to answer those questions. But my guess is play probably does help brain development. That makes me worry about all the computer games kids play today. I don't know whether the overall amount of locomotor play American children do has gone down, but if it has that's probably not good. When I was a child we didn't have game systems or computers or cable TV; we had two recesses a day at school instead of just one; and the only time of the week when kids could watch cartoons was Saturday morning. To me it seems like we probably did more locomotor play if only because we didn't have anything else to do. If locomotor play is important to developing the brain, I wonder whether children today are getting enough of it.

This is a bigger question than just whether or not kids grow up to be well-coordinated adults. Physical movement is probably the basis of a huge amount of academic, social, and emotional intelligence. A lot of major psychologists, including Jean Piaget, the Swiss psychologist who mapped out the stages of children's cognitive development, have said that movement is basic to learning, and I agree. My drafting students who've never learned to draw *physically,* holding a pencil in their hand and moving it across a piece of paper, can't draw at all on the computer. You have to learn to draw by hand first and then move to the computer. Virtual drawing isn't a substitute for the real thing. I've seen this over and over again. Piaget said children learn by physically manipulating objects and seeing how they work.

That's *movement*. So if kids aren't getting as much locomotor play now as they did in the past, that could be a problem not just for coordination but for learning.

Physical movement is probably what caused the brain to evolve in the first place, as a matter of fact. Dr. Rodolfo Llinas, a neuroscientist at UCLA who wrote *I of the Vortex: From Neurons to Self*, says the brain evolved because creatures needed a brain to help them move around without knocking into things.[29] He gives the sea squirt as the ultimate example of what having a brain is all about. The sea squirt is a primitive organism with about three hundred brain cells that starts out looking something like a tadpole, and ends up looking a little bit like a turnip. For the first day of its life it swims around until it finds a permanent spot to latch on to. Once it finds its spot, it doesn't move again for the rest of its life.

Here's the interesting part: while it is swimming it has a primitive nervous system, but once it becomes attached to an object it eats up its own brain. It also eats its own tail and tail muscles. Basically the sea squirt begins life as a kind of tadpole, with a tadpole-like brain, and then turns into an oyster-class creature. Since the sea squirt isn't going to move ever again, it doesn't need a brain.

Dr. Llinas's theory is that we have brains so we can move. If we didn't move we wouldn't need brains and we wouldn't have them. So I won't be surprised if Dr. Byers and Dr. Walker are right that one of the primary purposes of play is to develop the brain.

ANIMAL ROUGHHOUSING

No one knows exactly why young animals and humans play with their friends and siblings, either. We do know social play always means roughhousing, which has led a lot of behaviorists to reason that play fighting must teach animals how to win a fight when they're grown up. On the face of it that always sounded logical, because young males usually do more play fighting than young females, the same way adult males do more real fighting than adult females. Behaviorists figured the play fights were practice for the real thing.

But when researchers tried to establish a direct connection between roughhouse play and adult fighting in squirrel monkeys they didn't find any connection. The squirrel monkeys who played the most didn't win more fights as adults, and the monkeys who *won* the most play fights when they were young didn't necessarily win the most real fights when they were grown up. There was no correlation one way or the other. That doesn't disprove the hypothesis, but it doesn't support it, either.

Another interesting fact: play fighting is nothing like real fighting. A lot of the moves that happen in real fighting never happen at all in play fighting, and the ones that do, happen in a different sequence.

We also know that the brain circuits for aggression are separate from the brain circuits for play. Testosterone, which can increase aggression, either has no effect on play fighting or actually reduces it. Sometimes roughhousing play will *turn into* a real fight, but inside the brain rough play and real aggression are two different things.

The other piece of evidence that play fighting isn't about learning how to win is the fact that all animals both win *and* lose their play fights. No young animal ever wins all his play fights; if he did, nobody would play with him. When a juvenile animal is bigger, stronger, older, and more dominant than the younger animal he is play fighting with, the bigger animal will roll over on his back and lose on purpose a certain amount of the time. That's called *self-handicapping,* and all animals do it, maybe because if they didn't do it their smaller friends would stop playing with them. This is also called *role reversal,* because the winner and the loser reverse roles.

Role reversal is such a basic part of roughhouse play that animals do it when they play games like tug-of-war, too. A friend told me a story about her mixed-breed dog, when he was a year old and fully grown, playing with the four-month-old Labrador puppy next door. The new puppy was about a third his size, but Labradors are fearless and up for anything so she wasn't fazed by his size. The two dogs liked to play tug-of-war with a rope toy my friend had out on his terrace, but of course my friend's dog was so huge compared to the puppy that it was no contest. If he used all his strength he'd end up just whipping the puppy around the terrace like a Frisbee.

But that's not what happened. Pretty soon my friend noticed that the puppy was "winning" some of the tugs. First my friend's mutt would pull the puppy backward across the terrace, then the puppy would pull *him* backward a way. My friend said her dog was "keeping the puppy in the game," and I'm sure she's right.

Some behaviorists say that the fact that all animals self-handicap might mean that the purpose of play fighting isn't to teach animals how to win but to teach them how to win *and* lose. All animals probably need to know both the dominant and the subordinate role, because no animal starts out on top, and no animal who lives to old age ends up on top, either. Even a male who is going to end up as the alpha starts out young and vulnerable. He has to know how to do proper subordinate behaviors.

PLAY AND SURPRISE

Marek Spinka, an animal researcher in the Czech Republic, has created a general hypothesis of play in animals. His theory is that play teaches a young animal how to handle novelty and surprise, such as the shock of being knocked off balance or a surprise attack.

If Dr. Spinka is right, that would explain why play fighting is so different from real fighting, because a play fight has to be constantly surprising to teach the young fighters to respond to novelty. Dr. Spinka's theory also goes along with self-handicapping, since changing roles in the middle of a play fight means that the animals put themselves in roles they don't normally have. A normally dominant young animal puts himself in the subordinate role, and a normally subordinate young animal puts himself in the dominant role. That's a novel situation.

Dr. Spinka's theory is probably related to Dr. Llinas's research on the brain and movement. Dr. Llinas says that a brain has to do three things to allow its owner to move: it has to set goals (where do I want to move to?), it has to make predictions (if I move this way will I crash into that tree?), and it has to rapidly process tons of incoming sensory data to make sure its predictions are coming true and its owner is getting where he wants to go in one piece.

All of that is a pretty good description of what happens in almost any kind of play in young animals, whether it's locomotor or social or *object play,* which is playing with any kind of object, like a ball or a stick. One time I watched Red Dog playing with a plastic bag in the field next to Mark's house. It was a windy day, and she would pick up the bag, carry it upwind to the fence, then put it down on the ground where the wind would catch it and blow it across the field to the other side. She'd chase the bag the whole way across the field and then, when she got to the fence, she'd catch the bag and bring it back to the upwind side where she could put it back down so the game could start all over again. It's hard to see any reason for that game other than the fun of setting goals (I'm going to chase that bag across the field and catch it), making predictions (which way do I have to move to catch that bag?), and rapidly processing a lot of incoming sensory data from her race across the field. When you watch a young animal doing object play it really does look like they've *got* to be developing their basic brain functions in some way.

Social play has all of the same qualities. Mark likes to play a "go fishing" game with Red Dog where he takes a bullwhip and flips the tip out and lets Red Dog grab on to it. Then he says, "Oh, I'm going to reel in a big one!" That's a social game, and it's pure loco-motion. When you look at what young animals do when they're playing, and put that together with the fact that animals do the most physical play while the cerebellum is forming connections, I think we'll probably find out that play is an important way that a young animal develops its brain's ability to guide active movement.

CURIOUSLY AFRAID

So far, research is showing that the primal core emotions—*rage, prey chase drive, fear,* and *curiosity/interest/anticipation*—are handled by separate circuits in the brain. That doesn't mean that more than one circuit can't be turned on at the same time, or that one emotion can't trigger another.

A friend of mine tells a story about her six-month-old mixed-breed dog's reaction to her husband when he came home from a two-month research trip overseas. When the dog saw her husband he was overcome by terror and joy at the same time. He hit the floor in fear, crying and screaming, and at the same time he kept lifting his eyes up to the husband and frantically wagging his tail in greeting. Then he'd jerk his head back down and carry on screaming and cowering, all the while creeping along the floor *on his belly* toward the husband. My friend said it was exactly like the dog thought he was seeing a ghost. He was terrified and overjoyed in the same moment, seeing someone he thought he would never see again.

That's a clear case of an animal having two warring emotions at the same time, and it stands out because you see this so rarely. In real life, animals seem to feel emotions one at a time, with one important exception: the emotions of fear and curiosity. ESB research shows that curiosity and fear come from different circuits in the brain, and you can turn each one on separately through electrical stimulation without automatically turning on the other. But I have observed that prey animals often feel both emotions at the same time. I don't know whether predator animals also experience both fear and curiosity at the same time, but I expect they probably do.

I've already mentioned that cows will investigate scary new objects or people in their environment. If you stand still in their pasture they'll start to walk up to you because they're curious. But if you make even a tiny movement with your hand they'll jump right back, because they're also afraid. Then as soon as you stop moving, they'll resume the approach. When they get about four feet away they'll stretch their heads out as far as they can so they don't have to get any closer than they absolutely have to, and then their tongues will come out another eight inches so they can give you a good licking and sniffing. They're still scared, though, because any little rapid movement, like your hair or your jacket blowing in the wind, will frighten them off again.

This goes on for fifteen or twenty minutes tops, and then they get bored with you. I tell photographers, "You've got fifteen minutes to get your pictures." After that the cattle won't come up to you again, and they won't let you come up to them. They'll just move away if you try.

The way they act is so striking that I've had more than one person who didn't know anything about cattle go out to a pasture with me and say, "She acts like she's curiously afraid." That is a perfect description of how cows react to novel stimuli: curiously afraid. It's the only example of animals being ambivalent that I'm used to seeing as a matter of course.

BREEDING EMOTIONS

Apart from the ESB studies, another piece of important evidence that the core emotions each have their own separate circuits is the fact that you can use selective breeding to change one without changing the other. We know this from a quail study done in France by Jean-Michel Faure. Dr. Faure looked at two different genetically inherited emotions: fear and *social reinstatement*, which means the tendency for an animal to want to get up close to his buddies.[30]

They tested this by putting a group of quail in a cage at one end of a treadmill, and then putting one lone quail on the treadmill going in the opposite direction from the cage. The quail had to run against the moving treadmill belt to get back to the cage. They measured how hard the quail tried to get back to his friends.

They also measured each quail's fear level and then correlated fear with social reinstatement. Their first set of findings was what they predicted: high fear and high social reinstatement go together. The more fearful the bird, the harder he tried to get back to his group. You see that in all kinds of animals, including predator animals who don't need to stick together to be safe. Marmalade cats are high-fear for cats and they're also high-social. No one knows why, but it's true. They're super-affectionate; they'll eat up petting, much more so than other cats. But if you make a rapid movement a marmalade cat is the first to run away.

The next part of the experiment is really important. They used selective breeding to see if they could separate fear and social reinstatement—*and they could do it easily*. It was not hard at all to breed a high-fear quail who didn't care about getting to his buddies, or a high-social quail who wasn't afraid of anything. Even though in real life the two emotions go together, in the brain they're separate.

We have some evidence for this in people, too. Various studies have shown that positive and negative emotions are probably created by different chemical systems in the brain. That's not surprising, but what *is* surprising is the fact that positive and negative emotions aren't inversely related. If you use a medication like Paxil or Prozac to lower negative emotions in a normal person, you don't automatically raise his positive emotions. They're separate systems.[31] (This probably explains why people with bipolar disorder can have *mixed states*, when the person is excited and maybe even euphoric at the same time that he's highly irritable.)

Intentionally or unintentionally, humans often separate animal emotions that normally go together through selective breeding programs. For instance, take the idea of breeding an animal for low fear. That might sound like a good idea, because high fear levels can make an animal nervous, high-strung, and hard to manage. But fear is an important emotion, and a person or an animal with abnormally low fear levels can be dangerous. He's dangerous, because in nature fear rides herd on aggression. A dog with normal fear levels might want to get in a fight with a rival, but he's also scared of getting hurt and that slows him down. The dog who's fearless doesn't think twice.

You see that in humans, too. A fearful boy is a lot less likely to start fights than a fearless one. It's not that the fearful boy doesn't get mad; he does. Anger and fear are separate emotions, and a high-fear person or animal can feel as much anger as a low-fear person or animal. The difference is that fear keeps an angry person from going too far. There's some interesting research on this in men and women, too. Males get in more physical fights than females, but females have just as much anger, and in some studies show more indirect aggression, like gossiping about a person they don't like or excluding them from the group, than males. So far psychological research has found that the reason women have as much anger as men but don't get in as many physical fights is that they also have higher levels of fear in angry situations. Fear is a constraint on physical aggression.

People take a big risk when they try to breed less fearful dogs. They could end up with some very dangerous animals. On the other hand, so far we're getting away with it with Labrador retrievers. Labs are low-fear *and* low-aggression, which is something you don't see in nature. I'm sure this is because breeders have been selecting for lower levels of *both* emotions. At least, I hope that's what they're selecting for. But with Labs, too, breeders are starting to see some of the problems that kick in from the traits that we don't realize are genetically connected.

One of the problems comes from the fact that we're breeding for calm/calm/calm, and we're starting to get a Lab who's so calm he's abnormal. If you do something aggressive like grab him by both jowls he doesn't react. People are also breeding the startle out of Labs, so if a car backfires he won't jump and run off with the blind person he's supposed to be leading. That makes Labradors good with children, who can be rough and unpredictable.

Labs have low pain, too, although that may be a trait they've always had, since as working dogs in Newfoundland they had to jump into icy water to get fish out of fishing nets. You can still see that behavior in Labs today. A young Labrador puppy will jump in a little kids' wading pool and start pawing the water like crazy, like he's trying to catch the fish in there.

The problem with making a dog so calm is that you breed all the motivation out of them. I talked to a lady at a guide dog school, and she said some of the Labs are worthless because they don't pay attention. People are getting worried they're creating a dog they won't be able to train. Even worse, they're starting to see epilepsy in some of the dogs. You can end up with epilepsy no matter what brain trait you're over-selecting for. That's what happened to Springer spaniels, who now have *Springer rage*. They kept breeding them to look super-alert, and they ended up with a form of epilepsy that creates sudden, out-of-the-blue episodes of aggression.[32]

Genetically speaking, Labs are strange dogs: low-fear, low-aggression, and high-social. That's not a normal combination. And anytime you use selective breeding to create an animal who's really different from what nature created, you can end up with some nasty surprises. I think people should be much more careful and *aware* when they're overseeing animal breeding.

I don't want to leave the impression that I'm against Labrador retrievers, though. They're one of the best purebred dogs we have; they're good family dogs and good working dogs, too. I just want to make sure we keep them that way.

ANIMAL FRIENDS AND ANIMAL WELFARE

People who own and manage animals need to think about animal feelings, because animals have the same core emotions we do. Just keeping animals healthy and well fed isn't enough; we need to give animals enough social contact with other animals—and with humans in the case of cats and dogs—to live an emotionally normal life.

Animal mothers, and some animal fathers, love their babies; animal babies love their mothers (some love their fathers, too); and almost all animals have some form of friendship. Even seemingly unsocial animals like giraffes are turning out to have friendships now that people are studying their social structures more closely. A researcher named Meredith Bashaw at the Georgia Institute of Technology in Atlanta started researching giraffe friendship after two female giraffes got extremely upset when the male giraffe they'd lived with for nine years in the Atlanta Zoo was taken away. Neither female had ever mated with him, and from what the humans could see, the three giraffes hardly interacted much at all. So no one was expecting the females to react badly when the male was moved. But both females were horribly upset and started repetitively licking the fence, which is a sign of stress.

The reason no one knew giraffes had friends was that the field studies of giraffes from the 1970s had concluded that individual giraffes did not form close attachments to other giraffes. Ms. Bashaw says, "Giraffes just seemed to move about the plains of Africa like random molecules in your coffee cup." But after the female giraffes got so upset in Atlanta, Ms. Bashaw went to the San Diego Zoo where the giraffes are free to move around a ninety-acre park and she could watch whether some giraffes stuck closer together than others.

She found that giraffes have buddies just like every other social animal we know. A giraffe will spend 15 percent of its time grazing with its friend, and only 5 percent of its time grazing close to any other giraffe. Another animal expert who has studied giraffe friendships since the 1970s, Julian Fennessy at the University of Sydney, says that among Angolan giraffes, who live in the Namib Desert, particular females spend a half to a third of their time with their female friends.[33]

In any social grazing group you find some mother-daughter pairs, but you also find animals keeping company with other animals they aren't related to. Dr. Fennessy has also studied a group made up mostly of males, and the males have friendships, too. Animal researchers find animal friendships in most or all mammals. I don't know if montane voles form friendships (they may not), but at this point we believe that all or nearly all mammals—and possibly most or all birds—form friendships.

For people, solitary confinement is one of the worst punishments you can put them through, and it's no different for animals. Animals need friends and companions, and humans need to make sure they have them.

4. Animal Aggression

D og owners are usually horrified the first time they see their beloved pet kill a helpless little furry animal. I remember the day my good friend Tina saw her golden retriever Abbey kill a squirrel on the quad of the University of Illinois. Even though Tina was studying for a Ph.D. in animal behavior, she was still shocked when she saw her gentle dog finish off a squirrel like an expert assassin.

It's even more shocking when you see Lassie kill for what looks like the pure fun of it. My friend Dave, who always takes his seventy-pound half-shepherd–half-hound mix out for runs with him, was stunned when Max shot out after a groundhog one day, seized the animal by the neck, and then shook it violently until it was dead. The dog totally ignored my friend, who was racing after him shouting, "Drop it!"

Max knew perfectly well how to obey the command "Drop it" when he had a shoe in his mouth. But there was no way Max was dropping a live groundhog.

The most upsetting thing was that Max didn't have the slightest interest in actually *eating* his kill. He brought the dead groundhog over to Dave, dropped it at his feet, and beamed up at his master, obviously expecting Dave to be mightily impressed. In a way, Dave was. This was the dog he trusted to play gently with his two-year-old son, and he'd just watched Max turn into a vicious predator who couldn't be called off once the kill was underway.

After that, Dave said he started to wonder why people and dogs get along together at all. We have 60 million pet dogs in this country, all of them predators wired to kill—why aren't there *daily* newspaper reports of hideous fatal dog attacks on humans, instead of the actual number, which averages out to about fifteen a year, based on the years 1997 and 1998.[1] That's one dog out of every four million. It's tiny. If there were a disease that struck only one in every four million people, only seventy people in our whole country would have it. (Dogs kill people a lot less often than people kill people, that's for sure.)

I had another friend who told me the same story. Her kids were young when she adopted what she thought was a shepherd-Lab mixed-breed dog from a shelter. By the time he was fully grown it was obvious from his markings and behavior that the dog had more Rottweiler in him than anything else; he was a *very* dominant animal. While he was still a puppy he preferred to spend evenings alone in his crate, instead of getting up on the bed with the family to watch TV. That's typical of dominant dogs; they like their "space." A dominant dog doesn't interact when *you* want him to; he interacts when *he* wants to. He'll let you know when he's interested.

Even worse, people she and her dog met on the street were saying he looked like he had some pit bull in him, too. My friend didn't think she'd accidentally adopted a pit bull descendant, but she *was* a little upset that her grown dog looked and acted so much like a Rottweiler. A study published in September 2000 found that Rottweilers and pit-bull-type dogs are responsible for the vast majority of fatal dog bites, with Rottweilers being number one.[2] Some of that is due to the fact that Rottweilers have gotten so popular there are a lot more of them around. But that's not all of it. In 1997 and 1998 pit bulls and Rottweilers put together were responsible for 67 percent of all fatal dog bites, and there's no way Rottweilers and pit bulls together make up 67 percent of the total dog population in this country. Not even close. (For a number of reasons, including owners' rights to equal protection under the law, the authors of the study did not recommend laws banning either pit bulls or Rottweilers.)

Seeing how dominant her dog was, my friend started to wonder about the whole dog-human relationship. She'd grown up with dogs herself, but now that she was a parent she realized just how much trust we place in these animals. People trust dogs with their lives; people trust dogs with *their children's* lives. It's pretty incredible when you think about it. I don't think my friend needed to worry as much as she did, though. Rottweiler mixes probably aren't any more dangerous than any other type of dog, except for the handful of purebreds known for low aggression, like Labs. I say "probably" because there *is* some raw data available on the numbers of mixed-breed Rottweilers who have attacked and killed humans, but it's impossible to interpret because we don't know how many mixed-breed Rottweilers are in the dog population. Just eyeballing the numbers, it looks to me like Rottweiler mixes aren't any more dangerous than any other mutt. But I can't say for sure.

My friend figured she needed to teach the puppy to be more sociable, so she would pick him up out of his crate and put him on the bed with her and the kids. The puppy would stay put, but his way of playing was incredibly aggressive. My friend said his jaws would be snapping open and shut like a little alligator's. Even though she'd lived with dogs her entire life, she would watch this puppy snapping and snarling away and think to herself, "Why do I have this *animal* up here on the bed with my kids?"

Well, probably she shouldn't have had the puppy up on the bed, seeing as how the first thing any dog trainer will tell you is that a very dominant dog needs to be kept down *low*. A dominant dog should never be at eye level with a human! However, today the puppy is a sweet and good-natured adult dog whom neighbors and guests all like to visit. He's still dominant, and the family still has to remind him of the proper hierarchy (humans on top, dog on the bottom), but he is a cheerful and devoted member of the family.

How does this happen?

AGGRESSION IN THE BRAIN

To understand animal behavior you have to start from the brain and work outward. For years animal behaviorists didn't have this option, and researchers struggled to come up with definitive classifications of animal behavior. Animal aggression was especially difficult to categorize, if only because there's so much of it. Naturally, different researchers would come up with different lists of core aggressive behaviors. Some lists were longer; some shorter. One behaviorist might make a distinction between *intermale aggression* (the tendency of two males to fight when one male is dropped into the cage of another) and *territorial aggression* (which often means one male fighting another male who has invaded his territory, although females can engage in territorial aggression, too). Another researcher might decide that intermale aggression and territorial aggression were really the same thing.

Studying the brain doesn't solve all of these problems, because different behaviors can come out of the same brain circuits. But now that the brain circuits for aggression have been fully mapped out, the nature of animal and human aggression is a lot more clear.[3]

We know now that there are two core kinds of aggression: *predatory aggression* and *emotional* or *affective aggression*. Predatory aggression is chasing down and killing prey to eat; emotional aggression is everything else.

I'll start with predatory aggression.

THE KILLING BITE

Predatory aggression isn't just something predator animals do. Prey animals also have the neural circuits for predatory aggression in their brains, though these circuits don't get activated very often.

Research with rats, who are prey animals, shows that you can elicit a biting attack in some rats by stimulating the same part of the brain you would stimulate to elicit a biting attack in a predator animal like a cat.[4] Even though a rat rarely hunts prey in the wild, he has the innate, built-in capacity to do it. Jaak Panksepp, the author of *Affective Neuroscience,* says researchers haven't been able to turn on a biting attack in all the rats he's studied, just in the ones who have a naturally strong inclination to "approach and vigorously investigate potential prey objects such as mice."[5] Still, these are perfectly normal rats, so the fact that you can produce a biting attack in especially aggressive rats means that the neural circuits are there for all rats; they just don't use them. The predatory chase drive is almost certainly present in all animals as a *potential* behavior.

The actual moment of the kill, called the *killing bite,* is a hardwired behavioral sequence that never changes. Each individual member of a species is born knowing how to perform the killing bite, and each individual member of a species performs the killing bite the same way. A Labrador retriever killing a groundhog will look exactly like a German shepherd killing a groundhog. In the laboratory you can turn on the killing bite by implanting electrodes into the predatory circuits in the brain and stimulating them with electricity. The animal doesn't have to be hungry, and no prey has to be in sight.

All predators have a hardwired killing bite, but the bite can differ from species to species. Dogs and cats bite down and then shake their prey to death; large cats such as lions, who kill large prey animals like antelope, often bite the animal's neck and then hold on until it dies of suffocation. They do that because an antelope is too big to shake to death. Usually when a predator kills his prey you don't see any blood. The dead animal looks perfectly intact.

Scientists call hardwired behavior sequences like the killing bite fixed action patterns because the sequence of behaviors is always the same. Fixed action patterns are turned on by *sign stimuli* or *releasers*. For *all* predators, rapid movement is a releaser that turns on predatory chasing and biting. Over the years I've read various reports where a person has been injured or killed by a tame lion or tiger. In almost all of these accidents, the cause was rapid movement. The person who was bitten fell down, suddenly bent over, or dropped a tool, and the sudden movement triggered the predatory fixed action pattern. I'm sure that's where the line "Don't make any sudden moves" comes from in police shows. Humans have the same built-in primitive reaction to movement, and in a tense situation a sudden movement can trigger a person holding a weapon to use it.

While the fixed action pattern is always the same, emotions can differ from one animal to another within the same species. If Dave had two dogs instead of just one, he might find that one of his dogs was more motivated to hunt down and kill a groundhog than the other. The briefest glimpse of a groundhog might trigger the chase and kill in one dog; the other dog might ignore the groundhog unless the animal was repeatedly shoved in his face. Both dogs would perform the actual kill exactly the same, but their motivation to get to that point could differ.

SCHOOL FOR HUNTERS

Predatory killing raises the question of how much animal behavior is learned and how much is instinctual. The answer is that it depends on the species. Animals with large, complex brains like a chimpanzee rely on learning much more than simple-brained animals like lizards. Dogs, cats, horses, and cows are somewhere in the middle. Their brains aren't as complex as a person's or a chimpanzee's, but are a lot more complex than a lizard's or a chicken's. So dogs and cats are more dependent on learning than chickens are, but they probably use more hardwired behavior than a chimp.

The next thing to know is that there is a difference between the fixed action pattern itself and the emotions that motivate and drive the fixed action pattern. The *emotion* of chasing down prey and the *behavior* of killing the prey are controlled by different circuits in the brain.[6]

Seeing the word "emotions" in this context might be surprising. Animal experts used to talk about *instincts,* which are the fixed action patterns, and *drives,* which we defined as built-in urges that made animals and humans seek the core necessities of life like food and sex. Instincts and drives described animal and human behavior well from the outside, but the concept of a drive didn't hold up well once researchers started mapping the brain. It was too broad and abstract, and when researchers looked for single, unified brain circuits underlying specific drives, they didn't find them.[7]

Instead of finding one unified circuit for a *hunger drive,* for example, they found two different circuits, one for the physical aspects of hunger, the other for the emotional. The physical aspects of hunger, called *bodily need states,* are things like low blood sugar, which signal that an animal needs something to eat. There's a separate circuit in the brain that handles bodily need states. But a bodily need state on its own isn't enough, which should be obvious to anyone who's ever known a person with anorexia. People and animals also need the emotion of SEEKING, which I talked about in the last chapter, to motivate them to go out and hunt or gather the food their body needs.

Researchers don't know exactly how a bodily need like hunger hooks up to the emotions of hunting; that's one of the questions people are studying now. But they do believe that virtually *everything* people and animals do is driven by some kind of feeling. We know how important feeling is from ESB studies of animal brains, and also from close study of human patients who have had brain damage. Antonio Damasio, whose book *Descartes' Error* has been extremely influential, has studied people whose emotions have become disconnected from their reasoning and decision-making processes. These patients can't even decide what restaurant to go to for dinner, even though they're hungry and need to eat. Emotion and hunger are separate circuits in the brain, and both need to be working.[8]

To sum up: fixed action patterns are built-in, brain-based behaviors that are always the same in every individual in a species. Emotions are built-in, brain-based *motivators* that vary in intensity and probably in frequency of expression from individual to individual. You'll still hear some animal researchers talking about drives once in a while, and that's not wrong if you're only describing an animal's behavior from the outside. It's just that the broad concept of a *hunger drive* or a *sex drive* doesn't correspond to the specific brain circuits that are turned on when people and animals seek food or love. More than one different circuit in the brain is always involved.

That gets us to what is *not* biologically fixed in the brain. Emotions are built into the brain, but everything an animal does to act on his emotions, *except* for the fixed action pattern, is learned. A dog is born knowing how to kill a groundhog, but he isn't born knowing that a groundhog is food. Strange as it may sound, a dog has to learn from other dogs that groundhogs are good to eat.

Predators have to learn from other animals *whom* to direct their hardwired predatory behavior against. If a puppy grew up in a house with a pet groundhog, the puppy would learn that a groundhog is not prey and would probably never attack it. That's why puppies need to be raised around toddlers, or at least exposed to them. Toddlers do the same kind of sudden, rapid movement prey animals do, so it's easy for them to trigger a dog's predatory killing behavior. Puppies have to be taught that toddlers are not prey.

It's not hard to teach a dog what's prey and what's not; you just have to make sure you do it. When I was little our family had a golden retriever who was a vicious cat killer. Ronnie was the sweetest dog around little children. I can remember trying to ride on Ronnie's back when I was about four, and he never even protested. But whenever he saw a cat he became wildly excited and would instantly chase and kill it. Ronnie had been thoroughly exposed to toddlers as a puppy, and he knew toddlers weren't for killing. But he hadn't been exposed to cats and had concluded that cats *were* for killing. He never got confused about the categories, either, because a dog is emotionally wired to learn *prey* and *not prey*.

Having to learn what to eat and what not to eat gives animals and humans the flexibility to adapt. If an animal had to rely on instinct alone to feed himself he would starve if his usual source of food suddenly disappeared or went into decline. He wouldn't be able to imitate other animals, either.

IS IT FUN TO KILL A GROUNDHOG?

The answer is yes.

First of all, behaviorists call predatory killing the *quiet bite* because predatory killing is *not* done in a state of rage. We know from brain research that during a kill the rage circuits in the brain are not activated, and we know from observation that the killer is always quiet. Killing bites are nothing like the kind of loud, screaming fights you'll see two animals from the same species get into. During territorial fights the rage circuits can be turned on, and a rage-filled attacking animal makes a lot of noise. But when a predator is on the kill, he just bites down hard and then shakes his prey to death.

Dave's impression that Max enjoyed killing the groundhog was right. We know this from the ESB studies I mentioned in the last chapter. Animals like having their predatory killing circuits turned on, and will turn them on themselves if you show them how. When you think about what predatory killing is all about, of course it ought to feel good, because predatory killing means dinner. Killing a mouse feels good to a cat the same way finding a luscious ripe banana feels good to a primate.

According to Jaak Panksepp, ESB studies show that predatory killing comes from "essentially the same brain areas" as the SEEK-ING circuit, which produces the pleasurable feelings of engaged curiosity, intense interest, and eager anticipation I mentioned in the last chapter.[9] When the SEEKING circuit is turned on, animals and people seek the things they need and want, like food and shelter, or a perfect pants suit at a department store or an advanced degree in physics. People and animals love the hunt.

But angry aggression feels *bad*. Animals and people do not like having their rage circuits turned on, and will avoid it if they can. Rage is a painful emotion. Inside the brain, predatory killing and angry aggression are not the same thing. Not even close.

THE HAPPY HUNTER

Anyone who has ever watched a dog kill an animal will tell you that the dog sure looked happy afterward. But since most people aren't going to get the chance to watch a dog kill a groundhog, if you really want to get a good look at an animal enjoying the hunt, spend some time with a cat. Cats are the super-predators of domestic animals. They can get especially carried away chasing, batting at, and pouncing on a red laser "mouse." Laser mice are a variant on the battery-operated laser pointers lecturers use in large lecture halls to point to an overhead screen. If you've never seen one, a laser pointer projects a tiny red dot that the lecturer can shine onto the part of the overhead screen they're referring to. In a laser mouse the dot is shaped like a mouse. The mouse shape is just a marketing tool; any cat who will chase a laser mouse will also chase a laser dot.

Some cats get so excited chasing the dot that they've been known to break their own bones or dislocate joints. One time I was at my friend Rosalie's apartment in New York and I was amazed at the way her two cats, Lilly and Harley, chased a laser mouse. You could lead Lilly and Harley around the whole apartment at a dead run, jump them up on the counter, back down on the floor, up a bookshelf— you could shoot them wherever you wanted them to go. They were so frenzied I had to be careful not to suddenly reverse the motion, because I could throw Lilly into a back flip, she was so focused on that dot.

I've never seen a domestic cat chase any other toy that way. I've also never seen a cat behave that way outdoors, chasing live prey. Lilly and Harley had gone into what behaviorists call *hyper-activation of the predatory chasing instinct*; they were so mindlessly fixated they could have injured themselves. I think that happens with laser pointers because cats can see the dot but can't catch it. Even when a cat puts his paws on the dot he can't feel it or hold it. The laser dot probably becomes a *super-stimulus* that keeps on stimulating the chase because the cat can't complete the sequence of chase and catch, so the chase instinct can't get turned off.

I was intrigued to find that even when I held the dot still, which I assumed would turn off the chase behavior, they didn't calm down at all but kept frantically batting and pawing at it on the floor. They didn't look like they were playing with the dot, the way a cat will play with its prey; they looked like they were still in chase mode. I suspect that the reason Lilly and Harley stayed so fixated over a motionless red dot was that the slight tremor of my hand was making the dot vibrate enough to keep them hooked in. I was holding my hand as still as I could, but the tiny movements of mouse-dot on the floor were enough to keep them going. That's how hyper-activated they were.

I've been told that some cats *don't* chase laser pointers, which is interesting. I wonder whether those cats may know more about hunting and catching live prey than indoor cats like Lilly and Harley do. Lilly and Harley aren't allowed outside and were never taught to hunt by their mother, whereas a cat with a normal outdoor upbringing learns what to chase and when. Outdoor cats also learn to inhibit their chasing instinct so they can stalk their prey and get close enough to catch it.

An outdoor cat who's learned all these things may not be interested in a laser mouse for a couple of different reasons. Number one, a laser dot is not food and they've made the connection between chasing and eating; and number two, the cat knows how to suppress his chasing instinct. He isn't a slave to rapid motion the way Lilly and Harley are. Whatever the explanation, the fact that some cats don't chase laser pointers, while others chase them so frantically they risk injuring themselves, shows you that what an animal chases is learned, not instinctual.

The cats' fixation on the dot reminded me of autistic fixations. It was totally mindless; nothing else in the world existed. Their whole world was a little dot. I was like that when I was a child. I remember dribbling sand through my hands and the rest of the world disappeared. I was hypnotized by the tiny reflections coming off each little grain of sand. I couldn't stop looking. Sometimes I would stare at falling sand on purpose, just to shut out overwhelming stimuli from my environment.

I think I was probably tapping into the part of the same prey chase circuit Lilly and Harley had activated. Like the cats, I was attracted to erratic movement, because it was the constant *changing* movement of the reflections that held my attention. The autistic brain, like all brains, seems to be attracted to rapid erratic movement. The difference is, we get stuck in it. Flags are another moving object that used to fascinate me, and I wonder whether some autistic children's love of rotating fans falls into this category, too. I didn't care about fans myself, but the movement of fan blades didn't look erratic to me. The autistic kids who really love fans are usually lower-functioning, and their visual processing may be more piecemeal. Maybe to some autistic children the little light reflections off of fan blades does look erratic, so they get hooked.

How Animals Manage Predatory Aggression

In the wild, tigers and other animals who hunt for food can't act like Lilly and Harley or they wouldn't survive. First of all, no wild animal has an unlimited food supply. A predator who chased and killed everything that moved would quickly run out of food.

Another reason why an animal living in the wild has to show some restraint is that he can't afford to waste calories on a chase that doesn't end in a meal. If he killed animals he wasn't going to eat, he'd then have to kill even more animals to replace the calories he used up chasing down and killing prey for sport.

Last but not least, mindless chasing like Lilly's and Harley's would make an animal less likely to catch prey, not more, because it short-circuits intelligent stalking behavior. Cats stalk their prey to get in the best possible position to pounce and catch it. That's the whole point. A cat wants to *catch* the mouse, not chase it in circles forever the way Lilly and Harley were doing. So predators have to be able to inhibit the impulse to give chase until they're in the best position to catch the animal they're after.

What all of this means is that an animal has to be able to inhibit his chase sequence, and he has to learn how and when to do this from other animals.

We know this is true from the behavior of animals raised in captivity and reintroduced to the wild. The television show *Living with Tigers* had a terrific episode about two cubs who had been raised by humans and then returned to the wild. At first they chased everything they saw, whether they were hungry or not. One night they killed seven antelopes in an orgy of predatory killing. It was like Lilly and Harley chasing the laser. They just kept chasing and kill-biting every animal that moved, one animal after another. They didn't eat them; they just killed them. The humans finally began holding them back, trying to teach them just to kill what they needed to eat.

The humans also had to teach them *what* to eat. When the young cubs were presented with a dead zebra they instantly performed a killing bite to the neck. I'm not sure why the cubs did that since obviously the zebra wasn't moving, but it may have been because the zebra was down on the ground. Maybe that was the trigger.

But after they performed the killing bite they made no attempt to eat the zebra. They didn't know the zebra was food; they thought food was something that came in the back of a truck. It was the same problem Dave's dog had with the groundhog. Nobody ever told him that a groundhog is meat. The humans had to teach the cubs that the animals they were chasing were also good for eating, which they did by cutting open the dead bodies and exposing the entrails.

The film footage of those tiger cubs is a good lesson on what a fixed action pattern looks like, and on exactly how far a fixed action pattern will take an animal in life. The tiger cubs were born knowing how to perform a killing bite, but that was it. The rest they had to learn. I assume that a normal animal learns from his mother and/or his peers to kill only what he intends to eat, though I don't know this for a fact. We do know, however, that almost no animal *routinely* kills prey animals on an indiscriminate basis.

The only wild animal I've seen who will sometimes violate this rule is the coyote. Most of the time a coyote eats the animals he kills, but occasionally coyotes will go on a lamb-killing spree, killing twenty and eating only one. I believe it's possible coyotes have lost some of their economy of behavior by living in close proximity to humans and overabundant food supplies. A coyote that kills twenty lambs and eats only one isn't going to have to trek a hundred miles to find more lambs next week. Any sheep rancher will have several hundred other lambs that will be just as easy to catch later on, and the coyote knows it. Wild coyotes have probably lost the knowledge that you shouldn't waste food or energy.

AFFECTIVE AGGRESSION

Affective aggression is completely different from predatory aggression. Affective aggression is *hot* aggression; it's aggression driven by rage. Compared to predatory aggression, in affective aggression an animal's emotions are different, his behavior is different, and his body is different.

A cat whose rage circuits have been electrically stimulated assumes an aggressive posture and hisses, and his hair stands on end (that's called *piloerection,* for erection of the hair follicles). His body is aroused. His heart beats faster, and his adrenal system kicks in. Stimulate the same cat's predatory circuits and his body stays calm. Jaak Pansksepp says you see "methodical stalking and well directed pouncing,"[9] with no increase in stress hormones. Humans have tended to mix up these two states, because the outcome is the same: a smaller, weaker animal ends up *dead*. But predatory aggression and *rage aggression* couldn't be more different for the aggressor.

Animal behaviorists usually classify the different types of rage aggression by the stimulus that triggers the aggression, and different experts have come up with slightly different lists.

This is mine:

1. Assertive aggression. This category includes dominance aggression and territorial aggression.
2. Fear-driven aggression. This includes maternal aggression to protect young.

3. Pain-based aggression.
4. Intermale aggression. Intermale aggression is influenced by testosterone levels.
5. Irritable or stress-induced aggression. This includes *redirected aggression,* such as when a cat gets agitated by the sight of another cat outside but can't get to it and so attacks another cat or person inside the house instead.
6. Mixed aggression. For instance, fear combined with assertive aggression.
7. Pathological aggression.

Assertive Aggression

Assertive aggression includes both dominance aggression—one animal attacking another to assert or maintain his dominance in the hierarchy—and territorial aggression, which is when an animal attacks to protect his territory from intruders. Assertive aggression is probably connected to the neurotransmitter serotonin in a fairly straightforward way; the lower the serotonin, the more aggressive the animal. Antidepressants like Prozac that increase serotonin levels can reduce dominance aggression in a pet.

Unfortunately, the connection between serotonin, assertive aggression, and actual *social dominance* or *alpha ranking* within the group still has to be sorted out. There is strong evidence from colonies of vervet monkeys that the dominant animal has the *highest* levels of serotonin and the *lowest* levels of overall aggression.[10] The lowest-ranking animals show the most random, impulsive aggression, while the leaders are calm and collected and get aggressive only when they have to defend the group.

We know this from Michael Raleigh's famous study of twelve vervet monkey colonies. He and his team removed the dominant monkey, always a male, from all twelve colonies, then gave a medication that raised serotonin levels to one of the two remaining males in the colony, and a medication that lowered serotonin levels to the other. That gave them twelve subordinate males in twelve different troops who now had higher levels of serotonin than they did before, and twelve subordinate males who now had lower levels.

Every single one of the subordinate monkeys whose serotonin levels had been raised became the dominant monkey of the pair. Then, when they reversed the medications, raising the serotonin levels of the monkeys who had previously had their levels artificially lowered, *those* monkeys became dominant.

The reason this whole area is so confusing is that we're talking about two completely different fields of research. We don't know whether the people who study dominance aggression in dogs are talking about the same thing Michael Raleigh was studying in vervet monkeys. So for the time being we have to make do with the standard definition of assertive aggression I'm using here.

Fear-Driven Aggression

Fear-driven aggression causes so much violence and destruction in the animal and human worlds that I've often asked myself, What is rage for?

Why do we have rage circuits at all?

When you look at animals living in the wild, the answer is simple. Rage is about survival, at the most basic brute level. Rage is the emotion that drives the lion being gored to death by the buffalo to fight back; rage drives a zebra being caught by a lion to make one last-ditch effort to escape. I once saw a videotape of a domestic beef cow kicking the living daylights out of an attacking lion. It was some of the hardest kicking I have ever seen. Rage is the ultimate defense all animals draw upon when their lives are in mortal danger.

When it comes to human safety in the presence of animals, fear cuts two ways. Fear can inhibit an animal or a person from attacking, and very often does. Among humans, the most vicious murderers are people who have abnormally *low* fear. Fear protects you when you're under attack, and keeps *you* from becoming an attacker yourself.

But fear can also *cause* a terrified animal to attack, where a less-fearful animal wouldn't. A cornered animal can be extremely aggressive; that's where we get the saying about not getting someone's "back up against a wall." An animal with his back up against a wall is in fear for its life and will feel he has no choice but to attack.

On average, prey species animals like horses and cattle show more fear-based aggression than predatory animals such as dogs. That shouldn't be a surprise, since prey animals spend a lot more time being scared.

I categorize maternal aggression differently from some researchers; I put it in the fear department. I think maternal aggression is fear-driven at heart because over the years I've observed that the high-strung nervous animals will *always* fight more vigorously to protect her young than will a laid-back, calm animal like a Holstein dairy cow. Many a rancher has told me that the most hotheaded, nervous cow in the herd is the one who is most protective of her calf.

Any mother, nervous or calm, will fight to protect her baby. That's why on farms the human parents always warn their children to stay away from mama animals. But the fact that it's always the most nervous, fearful mother who shows the most maternal aggression makes me think that maternal aggression is driven by fear, even when the animal is calm by nature. When mother animals think their babies are in danger, they feel fear, and their fear leads them to attack. That's my conclusion.

This brings me to the fundamental question you have to ask yourself any time you're trying to solve a problem with aggression: is the aggression coming from fear or dominance? That's important, because punishment will make a fearful animal worse, whereas punishment may be necessary to curb assertive aggression.

Pain-Based Aggression

This one is simple and is something all humans have experienced themselves. Pain makes you mad. A person in pain will become irritable and start snapping at the people around him, but an animal can easily become aggressive. Vets have to watch out for pain-based aggression with any animal who is suffering. A dog who has been hit by a car may lash out and bite its owner due to pain. An animal who has arthritis or some other painful condition may become aggressive when the painful limb or joint is manipulated.

Intermale Aggression

Intermale aggression is linked to testosterone levels, which is why castrating a male dog can stop his fighting other male dogs. However, castration doesn't fix *dominance aggression* in a dog, which leads Dr. Panksepp to believe that intermale aggression may actually be a third form of primary aggression separate and distinct from either predatory aggression or affective aggression. Time will tell.

Irritable or Stress-Induced Aggression

Animals who live in highly stressful conditions are more prone to aggression than animals living in reasonably calm conditions. I heard about an awful case of stress-induced aggression where a Border collie ate all her puppies. Borders are a nervous, high-strung breed, and this particular collie ate her puppies after she had been taken on a long car trip and brought to a new house. Her stress levels were already very high because she lived in a dysfunctional household that included a hyperactive teenager who could never sit still, and apparently the long journey and brand-new surroundings tipped her into violent aggression against her own pups.

Even a constant relatively minor irritant like a flea infestation can trigger stress-induced aggression in an animal.

Mixed Aggression

In real life animals probably experience more than one motivator for aggression pretty often. In particular, we know that fear-based aggression and assertive aggression often co-occur in dogs. Dr. Panksepp thinks this probably happens with maternal aggression in some cases, where the mother attacks out of fear *and* out of territorial aggression. He also thinks that if intermale aggression does prove to be a distinct form of aggression, separate from the rage circuits in the brain, it probably doesn't occur in its "pure" form very often. Two males may go into a fight eagerly, like two boxers ready for the championship match, but rage probably kicks in as one or both males start to feel frightened, frustrated, or in pain. Then you have intermale aggression mingled with potentially three different kinds of affective aggression.

Pathological Aggression

Medical conditions like epilepsy or head injury can produce pathological aggression in an animal. This is true in people, too. For instance, we know that a lot of prisoners who have committed violent crimes have had head injuries at some point in their lives.

GENETIC TENDENCIES TO AGGRESSION

Some animals are genetically disposed to higher levels of aggression than others no matter what the circumstances. There are bloodlines of rare horses that have killed or injured grooms, and cattle breeders have observed that certain genetic lines of bulls are more aggressive than others. I've already mentioned the behavioral problems that crop up with single-trait breeding. The rapist roosters are the most dramatic case, but many pigs have become more innately aggressive, too. A study at Purdue University showed that pigs bred to be lean got into more fights than pigs from a fatter genetic line.

The genetics of aggression is an especially thorny issue with dogs. Most people don't want to believe that there are some breeds, like pit bulls and Rottweilers, that are more aggressive by nature. (Pit bulls aren't an established AKC breed.) Usually these folks have known or owned individual Rotties or pits who were sweet and good-natured, so they conclude that when a Rottweiler or a pit bull shows aggression the problem is the owner not the dog. But the statistics don't support this interpretation, although it's true that statistics on dog bites aren't hard and fast.

There are lots of problems with dog bite reports. For one thing, there are a few different kinds of dogs that are called pit bulls, including some purebreds like the American Staffordshire terrier and some mixed-breed dogs. Another problem: large dogs do more damage when they bite people, so they're probably overrepresented in the statistics. Also, lots of purebred owners fail to register their dogs with the AKC, so it's impossible to know exactly how many purebred Rottweilers there are in the country and compare that figure to the number of reported dog bites committed by Rottweilers.

Because dog population data is imprecise, no one can nail down *exactly* what each breed's "aggression quotient" is compared to other breeds. Still, you can get an overall picture of which breeds are most dangerous by looking at medical reports of dog bites. *On average* Rottweilers and pit bulls are so much more aggressive than other breeds that it's extremely unlikely bad owners alone could account for the higher rate of biting. And if you're looking only at anecdotal evidence, there are plenty of cases of nice, competent owners with vicious Rottweilers or pit bulls. Aggression isn't always the owner's fault. Writing about pit bulls, Nick Dodman says, "Originally bred for aggression and tenacity, pit bulls, if provoked, will bite hard and hang on, making them as potentially dangerous as a handgun without a safety lock. . . . they can become quite civilized, developing into loyal and entertaining companions. But the *potential* for trouble is always lurking somewhere, as a result of their genes and breeding."[11]

The Monks of New Skete, the famous trainers of German shepherds in upstate New York who wrote *The Art of Raising a Puppy,* say that every breed of dog has its *freak bloodlines* that produce dogs who are much more likely to be aggressive.[12] Some people have always bred dogs with enhanced aggressive behavior to serve as guard or police dogs; there are also drug dealers and other unsavory types who have deliberately bred very aggressive dogs either for protection or because they're part of the illegal dog fight scene. These dogs are like a gun with a hair trigger and no safety.

As I mentioned earlier, Rottweilers and pit bulls are the worst offenders now.[13] But before Rottweilers and pit bulls got so popular the most dangerous breed was the German shepherd, and Chow Chows show up in dog bite studies as having a much higher *rate* of biting per dog than other breeds.

The same study also found that male dogs are 6.2 times more likely to bite people than are female dogs, and intact males are 2.6 times more likely to bite people than neutered males.

Finally, there are some animals, including some dogs, who are just plain trouble. It's not their breed and it's not their owners. It's them. They're born that way, and they are bad, dangerous dogs.

If you're trying to buy or adopt a dog with the *absolute least* genetic proclivity to aggression, your best bet is probably a female, mixed-breed adult. However, it's really not necessary to be hyper-vigilant about the genetics of dog bites when you're choosing a pet. Serious dog bites are so rare that from 1979 to 1994 only .3 percent of the U.S. population got bitten badly enough to seek medical care. When you consider the fact that just about everyone in America who isn't living in a prison or a nursing home has fairly regular exposure to dogs, that's a very small number. You're better off thinking about how a particular breed of dog, or a mixed-breed dog, will fit into your life.

ANIMAL VIOLENCE

People who love animals often think of animals as being aggressive but not *violent*. Only humans, they'll tell you, commit rapes, murders, or wage wars.

But that turns out not to be true. Some chimpanzees actually fight what Jaak Panksepp calls *mini-wars*. This is organized, violent behavior. Two groups of males from rival troops will meet at the border between their territories and fight. So many chimpanzees die in these mini-wars that in a lot of places the ratio of adult females to males is two to one. Jane Goodall has talked about how upset she was to find out that her beloved chimps could do something so awful. War is not unique to human animals.

I've heard many stories of violent behavior in farm animals. A woman I met told me about an expensive ram she bought from a small hobby farm (that's a farm whose owner raises farm animals as a pastime, not a full-time business). The ram was perfectly tame and gentle around people, so she thought he was fine, and she put him out with her twenty ewes. The ewes had already been bred and were in the early stages of pregnancy so they didn't come into estrus. The ram smashed their sides in and killed them all.

Many animals can be horrifically violent for no reason, it seems, other than the sheer desire to kill and maybe even to torture. It took many, many years for people to finally realize that dolphins, for instance, aren't the benign, perpetually smiling sea creatures they look like to us. Instead, dolphins are big-brained animals who commit gang rape, brutal killings of dolphin "children," and the mass murder of porpoises. In her book *To Touch a Wild Dolphin,* Rachel Smolker writes that male dolphins stick together in gangs and will chase a female down and forcibly mate her. Female dolphins don't form groups the way male dolphins do.[14] Reading the book I found the similarity between dolphin gangs and human gangs creepy.

There was evidence that dolphins were killing babies and porpoises for years, but researchers just didn't see it. They kept thinking that the porpoises must have been killed by boats or fishing nets. Finally someone pulled a porpoise who had just been killed out of the sea and found tooth marks on its side that perfectly matched the teeth of a dolphin. Ben Wilson, a dolphin expert at the University of Aberdeen in Scotland, told the *New York Times* that when he realized it was the dolphins who were doing the killing, his reaction was, "Oh my God, the animals I've been studying for the last ten years are killing these porpoises."[15]

Animal experts always manage to make infanticide seem not so bad. The standard explanation is that adult males have evolved to kill babies in order to bring the mother into estrus so she can have *their* babies. That could be true, but when you put infanticide together with other animal violence you may start to wonder just how *evolutionary* it is for an adult male to kill a baby of his own species or even his own group. Is animal infanticide really what nature intended? Or is it, at least some of the time, an aberration of what nature intended?

A videotape about the predatory behavior of killer whales made me see animal aggression differently. The different pods had each developed a different killing specialty. Some pods killed tunas they stole from fishing lines; some killed seals; some didn't do a lot of active killing. They just swallowed the fish whole. One pod had even figured out how to kill penguins, bite a hole in one end of the bird, and then squeeze on the other end until the insides came out of the feather "wrapper" so they could eat them. It was like squeezing toothpaste out of a toothpaste tube.

But one pod had become killers for sport. The cameraman filmed the pod separating a baby whale calf of another whale species from its mothers and killing it. They crashed their bodies on top of it over and over again, pushing it underwater repeatedly until finally it drowned. It took them six or seven hours to kill the baby. Then they ate the tongue and nothing else. It was horrible.

The report didn't say whether the adults were males, but I expect they were. We do know that most of the violence seen in killer whales is done by adolescent males, just as it is in humans. Sociologists have found that boys and young men between the ages of fifteen and twenty-four are most likely to be engaged in violence compared to other age groups. That makes me think that the kind of killing those whales were doing *isn't* evolutionary. Maybe it's a negative side effect of immature brain development.

With dolphins, researchers have pretty much reached the conclusion that much of the killing they do serves no evolutionary purpose. Dolphins will slaughter hundreds of porpoises at a time. The only imaginable evolutionary reason for this would be if porpoises compete with dolphins for the same scarce resources, like food. But they don't. Porpoises eat different food than dolphins do. Killing a porpoise doesn't increase a dolphin's chances of surviving and reproducing in any way. The only conclusion is that dolphins kill porpoises because they want to.

I don't know why animal violence happens, but when I read through the research literature I'm struck by the fact that the animals with the most complex brains are also the ones who engage in some of the nastiest behavior. I suspect people and animals probably pay a price for having a complex brain. For one thing, in a complex brain there may be more opportunities for wiring mistakes that will lead to vicious behavior. Another possibility is that since a more complex brain provides greater flexibility of behavior, animals with complex brains become free to develop new behaviors that will be good, bad, or in between. Human beings are capable of great love and sacrifice, but they are also capable of profound cruelty. Maybe animals are, too.

WHY DOGS DON'T BITE PEOPLE

All animals have ways of managing their aggression. This is one place where evolution has to come in: it might be good for an individual animal to murder his rival, but it wouldn't be good for the species if it was normal for animals to fight each other to the death. Few adult animals apart from humans ever attack each other so violently that one of them dies.

Dogs have an inborn guard against excessive killing called *bite inhibition*. A typical dog learns bite inhibition through puppy play. Dr. Michael Fox of the Humane Society of the United States has found that prey killing and head-shaking movements first occur in four-to-five-week-old puppies during play, and if you watch two puppies playing it's incredibly violent. They'll snap and snarl and lunge at each other's throats—I've seen one puppy grab another puppy's throat, bite down, and shake his head violently, just like he'd do in a killing bite. But the minute the other puppy gives the tiniest squeak the biting puppy lets go. That's how they train each other that it's okay to bite "this hard but no harder." There are probably mechanisms to inhibit biting in all predators, because animals who are armed with teeth need to be able to stop biting before they rip each other apart.

Dogs have another method of teaching each other what's an acceptable level of aggression. When one puppy is getting too rough, the other puppy will suddenly stop dead in its tracks and stand stock-still facing the rough one. That always stops the other puppy, too. It's like a time-out. You'll see it a lot if you watch a younger, much smaller puppy roughhousing with an older, bigger puppy. They're both puppies, and they're both young, but one puppy is getting the worst of it thanks to size and age. It's amazing how fast the two puppies will adjust to each other's relative size and age. The smaller puppy will get lots rougher, and the larger puppy will get gentler.

Owners who play rough with their dogs are relying on their dog's bite inhibition to keep from getting mauled. Trainers say that's not a smart thing to do, because happy play can escalate to angry play if they get too aroused. That's one of the problems with having a multiple-dog household; the fun can turn violent and two playing dogs can suddenly bite each other for real. Still, even though all trainers tell owners not to play rough with their dogs, owners almost never listen, and I haven't read about people getting mauled by their pets in the middle of roughhouse play.

Roughhouse play is normal between dog friends, and it's probably normal between people and their dogs, too. I *have* seen people
play too roughly with their dogs, though. I saw an owner one time
play so roughly with his dog that it stopped being play to the dog,
and it made her yelp. He was grabbing her loose skin too hard, and
she finally growled at him. That's wrong.

I want to lay to rest one standard piece of dog trainer advice. Playing
tug-of-war is probably not as bad as people think. Most trainers will tell
you that playing tug-of-war with your dog encourages him to think
he's your equal, which is bad. Other trainers take a slightly different
view, which is that if you let your dog win a game of tug-of-war he'll be
less obedient, but if you win he'll be more submissive.

However, a study of fourteen golden retrievers in Great Britain a
couple of years ago found that neither of these things was true; at
least neither was true with the fourteen golden retrievers the experimenters tested. The researchers had people either win or lose a series
of tug-of-war games with the retrievers, and then watched how the
dogs behaved. The losers *were* more obedient after playing the
game—but so were the winners. All the dogs were more obedient
after playing tug-of-war with humans! And none of the dogs suddenly got more dominant. The winner dogs didn't display any dominance behaviors like raising their tails up high or trying to stand over
the person they'd beaten.[16] One study doesn't prove anything, but I
think it's probably both safe and fun to play tug-of-war with your
dog, and it might even be good for him. Just remember one thing:
the study also found that the dogs who lost every time were a lot
less interested in playing any more tug-of-war. Apparently a dog
doesn't like losing all the time any more than a person does.

THE BOAR POLICE

Pigs have a mechanism for managing their own aggression that I call
the *boar police*. Pigs can be really vicious. Any child raised on a farm
gets warned repeatedly to stay away from the mama pigs especially.
That's good advice, because pigs don't have a bite inhibition mechanism that I can see, possibly because pigs are more chewers than biters.
When I visit a pen of pigs they'll start nibbling on my boots; then grad-

ually they'll work up to chewing harder and harder until I say "Ouch!" They don't take a social cue like that, either. If the chewing starts to hurt I have to really get on them to make them stop.

When pigs *do* bite, it's bad. Fortunately, the presence of a mature dominant male in the group will inhibit fighting, something that's probably true of many other species, too, although it isn't well researched. We do know it's true of elephants. Marian Garai, a zoologist in South Africa, has observed aggressive behavior in young but fully mature bull elephants being kept under control by older dominant males.[17]

I did an experiment at a pig farm in Colorado to determine whether placing mature boars in a group of juvenile pigs would reduce fighting. Pigs can be nasty fighters and when strange pigs are mixed in together they will often injure each other as they fight to determine the new dominance hierarchy. I already knew this from earlier research by John McGlone at Texas Tech University.[18] He found that just spraying the scent of a mature boar reduced fighting, so I wanted to see what would happen if you actually put a live boar in there with them.

Having the mature boar present in the pen worked even better at controlling the fights than just the scent. With the boar there, both his scent and his behavior inhibited the younger pigs. When two pigs started a fight the boar would walk over toward them. That's all he did, just walked toward them. The only intervention was his commanding presence and his attention.

When the younger pigs saw him coming they stopped fighting. It was exactly like a bunch of young hoodlums who see the police and instantly stop what they're doing. The younger pigs acted so much like young human males they would even look around to see where the boar policeman was *before* starting a fight. If he was nearby they didn't start fighting, but if he was over in the other end of the pen they were more likely to attack.

SOCIALIZING ANIMALS TO OTHER ANIMALS

Anyone who interacts with animals has to know how to manage an animal's aggressive nature. Two actions are essential: make sure the animal is properly socialized *to other animals,* and make sure the animal is properly socialized *to people.*

You have to make sure animals are socialized to other animals, because most of what animals do in life they learn from other animals. Adults teach their young where to eat, what to eat, whom to socialize with, and whom to have sex with. The adults teach the young ones social rules and respect for their own kind. If an animal does not learn these rules when he's young, there may be many problem behaviors when he grows up.

One of the worst things you can do to any domestic animal is to rear it in isolation. Many people mistakenly believe that stallions are aggressive nutcases you can't handle, but that's true only because we make them that way. I remember being amazed when I walked into a holding pen at a Bureau of Land Management adoption center, which contained fifty wild stallions. The stallions were completely peaceful and quiet with almost no fighting. Every year the BLM gathers surplus wild horses and puts them up for adoption so that the horses don't over-graze the ranges, and people who visit the BLM pens find it hard to believe that fifty stallions can actually get along with one another. But that's the way well-socialized animals of any species usually behave. In the wild, constant fighting is not normal.

On the plains, subordinate stallions live together in bachelor groups. There's one dominant stallion who has all the mares to himself, like a harem; the rest of the stallions all band together and live in another group. The bachelor group tracks peacefully along with the harem group until the day when the dominant stallion has grown weak due to age or illness, and is ready to be replaced by a younger, stronger stallion. Only then does the younger stallion challenge him, not before.

Stallions would have to get along with one another to stay alive. Prey animals live in groups; that's how they survive. Wild horses in herds take turns sleeping and keeping watch for predators. If they had to live on their own they'd be killed in their sleep.

I mentioned this in the last chapter, and I'll say it again here: the modern fancy stable is a super-max prison for stallions. When a stallion is raised in solitary confinement he never learns normal social behavior, and that's what makes him dangerous to other males.

While they're growing up, young colts learn that there is a give-and-take to social interactions. They also learn exactly how horses establish and maintain a dominance hierarchy. All animals who live in groups—and that includes most mammals—form dominance hierarchies. It's universal. Researchers assume that dominance hierarchies evolved to keep the peace, because when each animal knows his place and sticks to it you have less fighting over food and mates.

No one ever knows for sure why one thing evolved and another didn't, but in the wild dominance hierarchies are usually stable once they've been established. Fighting levels drop and remain low until a new animal is introduced or an old dominant animal who has become weak is dethroned by a younger, stronger animal. If the animals in a dominance hierarchy are too evenly matched you might see a situation where no clear winner is able to emerge, so the animals keep fighting. That's not uncommon, but it's not the norm. Dominance hierarchies seem to minimize fighting.

Domestic animals are the same. Growing up with other horses, a young colt learns that once a stallion has achieved a certain position in the hierarchy he no longer has to keep kicking or biting the other horses. He also learns that no one challenges the dominant stallion unless he has a good chance of winning. Dominance hierarchies among horses are not like competitive sports in humans, where individual competitors or teams go head-to-head for the life of the athlete or the team. Subordinate horses don't keep on challenging the lead stallion day in and day out until somebody gets lucky and wins. They wait until the lead stallion is ready to be deposed. That's the rule.

But a horse isn't born knowing the rules; he has to be taught the rules by other horses. A stallion locked up in solitary confinement in a fancy show barn *is not normal.* He's especially likely to show abnormal aggression. There may be another reason for this, besides the fact that isolation-reared animals haven't learned proper social etiquette. Horses are social animals, and it's possible that a super-max stallion becomes a psycho fighter from emotional damage due to too much time spent alone. He might have more easily activated rage and fear circuits in the brain.

When I was in high school the myth that no stallion can get along with another male seemed true, because when my high school brought a fine big stallion named Rusty to the stable, all hell broke loose. Up to that point they'd had only mares and geldings, who all got along.

These horses were in a large field where there was plenty of room to move away from each other, but Rusty would charge wildly and bite and kick at the other horses. Soon it became obvious that Rusty could not get along with the others, so he was banished to solitary confinement in a pen between the horse barn and the dairy barn. Rusty had not been raised in a social group when he was a colt, and he was abnormally aggressive.

Raising young stud colts in a pasture full of older geldings will teach them some manners and create a good stallion that you can ride like a normal horse. People with fancy horses are actually abusing them with too much care. Young horses need to get out and have a chance to be horses.

It's not just stallions who can become aggressive if they're raised alone. A few years ago, I bought a piece of property on the west side of Fort Collins that has a thirty-acre horse pasture. Today my assistant, Mark, lives on the place and grazes his horses there. After the sale closed, I discovered that the big fat black gelding that was being boarded at my place had lived alone in my pasture for his entire life. Blackie was seven or eight years old, fully mature, and very gentle with people. He really liked to be petted, and I wanted to keep Blackie as a boarder.

But there was one major problem. Blackie was antisocial and he tried to kill every horse that was put in with him, male or female. On the thirty-acre pasture he would back a horse into a corner and kick it repeatedly with both back feet. I think that since Blackie had never learned any social skills he had never learned that once he had achieved dominance he no longer had to keep fighting.

After Mark moved into the farmhouse and brought his own horses with him, I learned that Blackie was now attacking Mark's horses. There was no way we could continue to board him, so Mark called the owner to come and take Blackie away.

Even cats are developing problems I think are due to isolated rearing. At the Colorado State University Veterinary Hospital there have been several "cat explosions" where the staff was severely bitten. I've actually seen that written on the charts: "Assistant was carrying cat down hall when cat exploded." This may be due to the cat leading a too sheltered life so kitty is seeing her first dog there at the veterinary clinic and she goes ballistic.

Julie, the lady who does my Web page, got a severely infected hand from one of these "fear kitties." She had adopted a friendly shy cat and one day when it saw a dog it instantly turned into super-fuzz-ball Halloween cat and bit her wrist to the bone. That cat needed to see some dogs at an early age so she could get used to them. But today fewer and fewer house cats are learning about dogs. Some animal shelters even make adoptive owners promise never to let their cats out of the house. That might keep the cat from getting run over, but what happens when you take him to the vet? Pet owners need to socialize new kittens and new puppies to other animals not long after they bring them home. If they reach adulthood without being exposed to other animals, it's probably too late.

I think dogs may be starting to have aggression problems due to overly isolated rearing, too. All of the leash laws towns have passed may be having some adverse effects on dog socialization, because unless the owner makes an effort, many dogs do not get properly socialized to other dogs, or to other people. We need these laws, because stray dogs running loose can be dangerous, especially if a group of stray dogs starts thinking of itself as a pack. Several dogs together are more dangerous than one dog on its own, because *pack mentality* can set in. But leash laws have probably had a cost.

When I was a child all the dogs ran loose in the neighborhood and there were very few dog fights (and almost no dog bites to humans) as a result. Our golden retriever Lannie was subordinate to Lightning, who lived next door. Lannie knew his place, and when Lightning came near he calmly rolled over in submission. I never saw Lightning bite him. All the neighborhood dogs were socialized to one another, and they knew their place in the hierarchy.

The breeds were Labrador, golden retriever, German shepherd, and mongrel. No pit bulls or Rottweilers. The scariest dog in the neighborhood was a Weimaraner, who went stir-crazy in his owner's house. Butch did not get enough exercise and he was absolutely hyper from being locked up all day alone in the house. Anytime you rang the doorbell Butch flung himself at the window beside the front door.

Butch turned out to be a killer of other dogs. One day Butch and the police department's German shepherd police dog were being walked in the park by their owners. Butch broke away and killed the policeman's dog. This is an unfortunate example of what can happen when a dog is not socialized to other dogs when he is young.

I'm a little concerned that leash laws may encourage dog-on-dog aggression even in dogs who *have* been well socialized to other dogs. One of my friends owns a highly dominant seventy-pound male mutt, and her next-door neighbor owns an eighty-pound male golden retriever who is also highly dominant. The two dogs played together throughout puppyhood, and were close friends. But as soon as their testosterone kicked in they began to fight, and they continued to fight even after both dogs had been neutered. They've had two battles now, both resulting in injuries bad enough that a veterinarian had to stitch them back up. Even worse, the owners had to break up both fights at high risk to themselves, because neither dog yielded. These are two well-socialized, well-cared-for, normal, healthy dogs who played together as puppies living next door to each other. And now they're trying to kill each other. That never happened in my neighborhood when I was growing up.

I should probably add that the fact that the mutt is just as aggressive as the golden retriever doesn't say anything about mutts versus purebreds, because the selection pressures on mutts should make them better socialized to *humans,* not to other dogs. This particular mutt is perfectly behaved with his human family and their friends and relatives. It's other dogs he has a problem with.

The reason I think leash laws may be part of the problem is that both dogs are kept inside their respective yards at all times. I'm guessing that leash laws may be short-circuiting some core principle of animal behavior in the wild. In nature, where animals are free to come and go, animals almost never seriously injure other animals who are familiar to them. But I've found that dogs living side by side in fenced yards often *do* hurt each other if they can, even when they've known each other for years. This may be a case where proper socialization won't help. The dogs *have* been properly socialized, but their environment—a fenced-in yard—is "improper."

ORPHANS AND OTHERS

Animal rescue programs also have had terrible problems with aggression, because the young animals they save are usually orphans. There've been horrible problems with orphaned elephants who did not have the opportunity to grow up with their own kind and learn proper elephant social ways. The males are the worst. When they grow up without an older experienced male to guide them, their behaviors become vicious and bizarre. Turning young orphan male elephants back into the wild has been a disaster. They will sometimes seek out rhinos and either kill them or try to mate with them. Their behavior is completely off-the-wall.

An animal who hasn't been properly socialized to his peers isn't dangerous only to other animals. He can be dangerous to humans, too. In social grazing animals such as horses, deer, and cattle, the hand-raised pet bull often becomes the most dangerous. The problem there is mistaken identity. A hand-raised bull calf thinks he's a person instead of a calf.

That's fine until he becomes sexually mature at age two, and instead of going out and fighting another bull to establish his dominance, he attacks the person who raised him. Bulls establish dominance by butting each other with their heads, and no human can survive being head-butted by a thousand-pound animal. It's essential that bull calves not get confused about their identity. They are cattle, not people.

Ranchers can prevent their cattle from identifying with humans by rearing calves with their mothers inside a herd of cattle. A study done by Ed Price at the University of California indicated that Hereford bull calves raised by their mothers almost never attacked people, but calves raised alone in small pens often attacked people when they grew up.[19]

When I visited Australia I heard a tragic story about a person who hand-raised a deer fawn to adulthood. One day when the

owner knelt down to photograph him, the deer interpreted the
man's kneeling down posture as the head-bowing behavior of
another bull challenging him. He charged and gored the owner to
death with his antlers. It's so important to raise calves with their
mothers. When a young bull calf or buck deer fawn is raised with his
own species he'll direct dominance attacks toward his own species
instead of toward people.

This will probably come as a surprise, but huge social animals like
cattle are actually more dangerous to handle than big solitary predators like tigers. A bull can attack a person to achieve dominance, but
a tiger won't, because a tiger doesn't care about dominance; constant jostling inside a social hierarchy just isn't part of a tiger's life.
You have to be extremely careful not to trigger predatory aggression
in any big cat, obviously, but that's all. Every year several ranchers
and dairymen are killed by cattle challengers, and it's my opinion
that the best way to prevent dangerous attacks on people is to raise
highly social grazing animals like cows and horses strictly with their
own kind. They should look up to people as a benevolent higher
power. You don't want a cow directing any cow aggression at
humans.

To help prevent attacks from orphaned male grazing animals, the
animal should either be fostered onto a new mother or penned
together with other young males. In both cases the young bull will
learn that he is a bull, not a person. It is also important that the
young bull be castrated at an *early* age. By early, I mean the animal
should be castrated before he has matured physically. (Dogs are usually castrated after they've reached physical maturity.) Castration will
greatly reduce aggression in grazing animals. If a bull calf is castrated
at a young age he can be safely raised in your backyard. That's why
kids in 4-H and FFA (Future Farmers of America) show thousands
of steers safely every year. They aren't raising bulls.

SOCIALIZING ANIMALS TO PEOPLE: DOGS

Domestic animals have to be socialized to people, too. We call dogs
man's best friend, but one and a half million dogs get put to sleep
every year because of behavior problems the owners can't live with.

A lot of those problems are dog bites. If you are going to get a dog, you can't plan on preventing dog bites by keeping your dog safely locked up in your house or yard, either, because dogs almost always bite people they know, usually people they know well. Around four and a half million people get bitten every year, and the Centers for Disease Control report that over 75 percent of the dogs in these incidents belong either to the family of the person who got bitten, or to a friend.[20]

Predator animals are built to hunt and kill, and they're less fearful than prey animals. That makes them *potentially* dangerous to people for two reasons: a person can accidentally trigger a predator's killing bite instincts through sudden movement, *and* a predator animal is less afraid of expressing angry aggression. Left to their own devices, dogs can become dangerous to other dogs, to cats, and to humans, and you can easily train a dog to be hideously ferocious if that's what you want. Dogs are so aggressive by nature that the Monks of New Skete say a trained guard dog is like a loaded gun, and families should not own them. Only a professional can live safely with a trained guard or police dog.

That fact alone tells you a lot about the difference between predator animals and prey animals. You couldn't train a horse to be an "attack horse" even if you tried, although a horse who feels threatened can be very dangerous. You can create an "attack animal" only out of a predator animal like a dog. So if you're going to own a dog, you need to teach him that it is unacceptable for a dog to threaten or bite a person.

It's especially important to socialize dogs to children. Most of the fatal dog bites involve young children, because they're low to the ground and they run around a lot. The dog mistakes the small running child for prey, and attacks. *All* predatory animals have to learn which animals are prey and which are not. A dog does not know that your two-year-old is not prey unless you specifically teach him this while he's still a puppy.

You also have to be careful to teach your puppy that *other people's two-year-olds* are also not prey. That's easy; you just have to make sure your puppy gets exposed to toddlers who don't live with you. Since a lot of toddlers love to run up to strange puppies and hug them, you can accomplish this by taking your puppy for walks in parks where parents bring their children to play, or in neighborhoods with lots of families. After your puppy has met a few little kids on outings, he'll know that small children are not prey. I want to stress that it's essential to introduce your dog to other children in other families, because to a dog, *your* two-year-old and *the neighbor's* two-year-old are two different categories; they're apples and oranges. A puppy doesn't automatically generalize don't-attack-Johnny to don't-attack-Joey.

KEEPING THE PEACE

This brings up the question of dominance. *All* animals who live in groups—and that is most mammals—form dominance hierarchies. Animals are not democratic and there is always an alpha animal, and often a beta animal, too. Dogs have an alpha male who is dominant over the others, as well as a beta male who is second in line to the alpha.

Dog owners must establish themselves as the alpha, period. *This is the one rule you must not ignore.* A dog who thinks he's the alpha in the house is dangerous, because dogs will fight any lower-ranked pack mate who challenges them. If the family dog becomes the alpha he's going to be especially dangerous around important resources like food and his resting place. He'll bite family members who come too near his dog dish or sit down too close to him on the sofa when he's taking a nap. He's definitely not going to cooperate with any trips to the vet, either.

This happens more often than you'd think. There are plenty of households where the dog is the alpha. You can't necessarily avoid the problem by getting a female dog, either. According to the American Veterinary Medical Association, 80 percent of dogs brought in to see vets because of dominance aggression are unneutered males, but you can definitely have dominance aggression in neutered males, and in females, too, whether they're neutered or not. Actually, when it comes to females, Nick Dodman says that an aggressive female can actually get worse after she's been spayed, because she doesn't have as much progesterone in her system to calm her down.

Even though unneutered males are the biggest biters, neutering a dog once he's *started* to bite probably won't solve the problem. With animals, there's a huge difference between preventing aggression in the first place and trying to stop it once it's developed. Dr. Dodman says that in his experience neutering a male doesn't make him any less dominant, or any less likely to bite a *human*. Neutering an aggressive dog mainly keeps him from biting other dogs—but not because he suddenly becomes submissive. Neutering probably decreases dog-on-dog aggression only because the neutered dog stops smelling like a male to other males, so other males don't challenge him as much. It's not that the neutered dog is any less dominant after he's been altered; it's that other male dogs are nicer to him.[21]

One of the most upsetting situations with an aggressive dog I've seen over the years happened to a family I knew with two very young boys and a father who wasn't very nice to the mom. The dad was always saying mean things to her in front of the boys and the dog. Then, when the boys were still little, the family broke up. The mom took the boys and the dog and moved to another state to start graduate school.

Not long after that the dog went crazy. He began threatening to bite the mom if she tried to pull him someplace by the collar, and he constantly tried to keep her from leaving the house. One registration day as she was getting into her car to go sign up for courses, the dog jumped in the back seat and wouldn't get out. He growled and snapped viciously at her face each time she tried to take hold of his collar to pull him out. He ended up sitting inside her car the whole day, until *he* decided it was time to get out. Things got so bad that the only way the mom could manage the dog at all was to trick him by throwing a piece of steak wherever she wanted him to go, then slamming the door behind him when he ran after it. Her friends were all frightened of the dog, and so was she.

Her boys weren't doing well, either, and the child psychologist she took them to see said that the dad had treated her so disrespectfully her sons didn't trust her to take care of them. They didn't think she could do it, and they were scared.

This was probably a case of disrespectful behavior inside the family affecting the dog as well as the children. The husband was probably the absolute alpha in the dog's eyes. The dog may even have concluded that he was the beta animal, because the wife was so downtrodden. So when the husband disappeared the dog immediately challenged the wife for alpha status. That is always a dangerous situation.

I lost track of the family not long after the registration day incident, so I don't know whether the mom was ever able to get the dog under control or not. Things had reached the point where she needed to hire a trainer, but I knew she couldn't afford it. I hope thinks worked out for them, but it didn't look good.

Establishing dominance over a dog is easy. Many people think that exerting dominance means beating an animal into submission, but that's not true at all. I am totally against rough *alpha rolling of dogs,* which is still used by some police departments to train police dogs. In alpha rolling a person throws a dog over on his back and holds him down. Rolling over and exposing the belly is a hardwired instinctual behavior in dogs, and a well-socialized adult dog usually rolls over on his back to be petted. That's why you want your puppy to spend some time on his back looking up at you; just being in that position reinforces the fact that he is subordinate to you.

But you shouldn't *force* him onto his back. When two dogs from the same pack meet, the subordinate dog will voluntarily roll over; the other dog doesn't shove him over. When a dog is forced into this position by a human the hardwired submission behavior does get turned on, but when the dog stands back up he does not forget having been forced down. Someday, when your back is turned, he will bite you in the butt.

A much better way to train the dog is to make rolling over a fun game for him through tickling or stroking his chest or belly and offering food treats when he rolls over. This makes the dog get into the position of submission without anything aversive being done.

I also want to say something about the overall issue of punishment in animal training. I am totally against using punishment to teach an animal new skills. In almost all cases animals can be trained to do tricks or develop skills using positive methods.

The one exception is stopping dangerous prey-drive-motivated chasing of joggers, bicyclists, and cars. In this situation a shock collar may be needed. If you do have to use a shock collar to stop your

dog from chasing people and cars, it is important that the dog never
figures out that it's the collar that gave him the shock, so you should
leave the collar on for a few days before using it. When your dog
receives a correction for chasing a jogger, you want him to think
that the dog god did it.

The best ways to establish dominance are obedience training and
making the dog sit quietly before he is fed. The dog should learn
that he eats on his owner's terms. You can also do things like going
inside the door first before letting your puppy enter, putting your
hand in his food dish while he's eating, and playfully coaxing and
rolling him over on his back (*not* throwing him over). Some trainers
even recommend growling at your puppy like a mama dog and nip-
ping him on the muzzle when you're giving him a correction. I
know that sounds dangerous, but with a puppy it's not.

You also have to do at least some obedience training. Obedience
training just means teaching your dog to obey a few commands. The
commands can be anything you like. You could get fancy and train
your dog to herd sheep, bring your slippers, or wear a tutu and spin
around in circles; it doesn't matter. The important thing is that the
dog learns to obey commands from his master.

You have to do obedience training no matter what your life is like.
Even if you live on a great big ranch where your dogs can run free
they still have to be obedience-trained, because your dog has to
know you're the boss or you are creating a potentially dangerous situation. That's the whole point of obedience training—*obedience*.
Not teaching your dog how to do tricks. Obedience training estab-
lishes the owner as the alpha.

It's amazing how easy it is to dominate a dog. When I was in college I went to visit a friend's house and they had a hound who had
become totally dominant. If Bernie wanted the softest chair, that
was the chair he was given. He was number one. He also had the
disgusting habit of lifting his leg and urinating on every guest.
Bernie was the king.

But there was one guest he never peed on and that was me. He
also never growled at me, or asked for my chair. Maybe it was my
posture and attitude, because I never did anything bad to that dog.
It shows how tuned in to people dogs are. That dog just *knew*, prob-

ably from watching me, that I would not tolerate being peed on, growled at, or any other obnoxious behavior.

PACK MENTALITY

Even when you establish yourself as the alpha, you can still have problems with other dogs, either with dogs in the neighborhood or other dogs in the household. Dogs need friends, and if you're going to be away at work all day I recommend owning two dogs, preferably a male and a female. But I'd stop at two, because more than two dogs in one house can be a big problem if the dogs are too evenly matched in size, age, and strength. With closely matched animals the dominance hierarchy may not stabilize, because no leader is able to emerge and the dogs continue to challenge each other. If you're going to have more than one dog the best plan is to stop at two, and to have one dog of each sex.

Another reason to stop at two is that dogs in a pack are much bolder and more aggressive than one dog on its own. *Pack mentality* is real. I mentioned the collie who pretends she doesn't notice the two barking German shepherds whenever her owner takes her for a walk. One day my friend took the collie and her other dog, the golden retriever, out for a walk along with the neighbor and *her* two dogs. The four dogs knew each other well, and probably felt like a pack.

This time the collie was a completely different dog. When they got to the German shepherds' yard and the two shepherds rushed the fence, the collie went nuts. She was slamming herself into the fence, barking, and racing back and forth from one end of the fence to the other chasing those dogs. She was really cussing them out, and it was all because *she was in her pack*.

She refused to leave, too. Her three friends got totally bored taunting those poor fenced-in dogs, and kept trying to get the collie's attention so they could go on with their walk, but the collie wouldn't budge. It was like she was making up for lost time. Her owner finally had to drag her away.

A dog pack can be incredibly dangerous to humans. A couple of years ago a ten-year-old girl in Wisconsin was killed by a pack of six Rottweilers while she was playing at her friend's house down the street. There were two adult dogs and four puppies in the house (which was a violation of a city ordinance limiting the number of dogs per household to three) and apparently the little girl began to pet one of the puppies, and one of the adult dogs got jealous and bit the girl. That set off the pack and they attacked.

Opinion varies on how to keep the peace if you do have more than two dogs. Most people, though not all, say you should always handle and pet the dominant dog first. The king must be treated like the king, although the ultimate leader is you. If you don't respect the dogs' natural hierarchy you can put the underdog in danger. Dr. Dodman has a horrible story about a pack of Chesapeake Bay retrievers living with a lady who indulged them totally and never gave them any obedience training. She lived alone, and the dogs were her surrogate family. Of course, in a real family children don't just naturally sit around behaving themselves nicely and saying "please" and "thank you." Children have to be obedience-trained, too.

The pack in this lady's house had formed a natural hierarchy, with two dominant dogs on top, two or three middle-ranking dogs, and two underdogs. But the lady refused to respect the ranking, and always lavished lots of time and attention on the two underdogs whenever she came home.

All that attention was provoking the top dogs into launching vicious attacks on the bottom dogs. Dr. Dodman told the lady she needed to greet and feed the dominant dogs first when she came home, but she wouldn't listen, and kept on showing favoritism. The whole thing ended in disaster. First one of the underdogs was badly injured and the lady decided the only way to deal with the situation was to have the little underdog put down. Then the one remaining underdog was horribly injured by the two top dogs and the lady had the two *top dogs* put down. All three dogs died just because this lady wouldn't listen to good advice.[22]

WORKING WITH THE ANIMAL'S NATURE: FARM ANIMALS

A human owner has the responsibility to understand and respect his pet's nature. *Dogs and cats are predator animals.* Dogs are hyper-social predators who live in dominance hierarchies. If you interfere with the hierarchy you can get the low-ranking dog or dogs killed by their own pack mates. You have to work with an animal's emotional makeup, not against it.

Domestic animals such as pigs, cattle, and horses are less controlled by purely social stimuli than dogs, so with these animals it's especially important to exert dominance the way another animal would do it. I learned this lesson when I was raising piglets as part of my Ph.D. work in animal behavior. My piglets lived in a Disneyland of straw with lots of different objects to root and tear up. I would sit with my piglets for hours and watch their behavior.

The one I named Mellow Pig would instantly roll over when her belly was scratched and would actively solicit people to rub her belly. But the largest pig in the pen did not like being petted at all, and she was the dominant boss hog. She thought she owned the place. Her coloration was what an Illinois farmer calls a blue butt; she had white forequarters and a grayish blue-gray rear. I named her Big Gilt.

When Big Gilt reached a hundred pounds, she started biting me whenever I entered the pen. The other pigs sought petting and stroking but Big Gilt disdained it. She just wanted to be boss. The bigger she got the worse the biting got and I had to stop it.

I tried waving my arms at her and shouting, but it didn't help. In desperation once I even tried slapping her big blue butt. That did no good, either. Finally I figured out that I had to act like a pig. I needed to assert my superior dominance by biting and pushing against the side of her neck the same way another, bigger pig would.

So, to simulate another pig biting and shoving against Big Gilt's neck I used a short piece of a one-by-four-inch board, about eighteen inches long, to poke and shove her against the fence. That's what the winner pig does: the winner pushes the loser away, or up against a wall. I shoved the end of the board repeatedly against her thick neck and I made it very clear that I was stronger, which I was. A full-grown human can still push around a hundred-pound hog. I didn't hurt her, but I did dominate her.

It worked like magic. Big Gilt stopped biting me and I was now Boss Hog. Using the hardwired instinctual behavior pattern was much more effective than slapping her. The only problem with this method is that it has to be done when the animal is young enough so you can still easily push the pig away. Again, I want to emphasize that I did not beat her up. She was overpowered by a stronger being who applied pressure to the right spot. Pushing the board against her neck turned on a hardwired instinctual submissive behavior.

After that Big Gilt was now polite when I entered the pen and she never bit me again, but she still did not like petting. One day while I was stroking Mellow Pig on the belly I started to rub Big Gilt on the belly, too. Since I was now the boss she didn't run away, but she clearly didn't like it. The strangest thing happened. Hardwired instinct collided with clear conscious will. Rubbing her belly triggered the instinctual rolling over behavior, but only the rear end of Big Gilt rolled over. Her front end remained standing when her hind end collapsed. The whole time I was stroking her a horrid growling sound came out of her throat. I had turned on the pleasure response to a belly rub, but the other end of Big Gilt did not want to give in. She did not dare bite me and she did not try to run away, but she surely did not like it.

PREVENT AGGRESSION IN THE FIRST PLACE

If I'd known more about animals I would have started establishing myself as Boss Hog a lot sooner, since as I mentioned earlier it's better to prevent aggressive behavior in the first place than to try to change it once it's developed.

Once an animal *has* developed aggressive behavior, in most cases it's going to be easier to deal with in prey animals than in predators. A good example is my friend Mark's horse, Sarah, who's nasty around the feed trough. Sarah was not reared alone, so she doesn't have the kinds of problems Blackie did. She's just got a bad attitude when it comes to food, and she'll chase away all the other horses so she can have the food to herself. I've seen a lot of horses do that.

All Mark has to do to deal with Sarah's nippiness is feed her last. Then, after she gets her food, if she still tries to run the other horses off he chases *her* off instead. It works like a charm for about two weeks. Sarah has perfect manners at the feed trough. Then she starts getting nasty again, and Mark repeats the procedure.

I talked to a vet student who has the same problem with a horse she owns, and she uses a slightly different version of the same technique. She doesn't feed *any* of the horses until they're all standing nicely at the trough, with their ears forward. Then she feeds them all at the same time. If anyone has their ears pinned back—any of the horses, not just the problem horse—no food. It's not hard to get the group of horses to turn their ears forward, because that's what horses do naturally when they're paying attention to you. She just waits them out, until all of them are focused on her instead of on each other. She uses Mark's approach only if her horse tries to chase the others away after the feed has been put out. Then that horse does get fed last. She said her system works really well.

The point is, you have to do a lot of emotional damage to a prey animal to turn it into a killer. As we've seen, if you lock a stallion up alone in a stall for his whole life, with no socialization at all, he could become aggressive. He might rear up and strike at people. That's dangerous, but only because the stallion is so big. He isn't actively trying to kill the person he's kicking. There are always exceptions, of course. Just recently I read a report about a stallion in Poland who became aroused by a nearby mare and then attacked and killed his owner, who was trying to calm him down. The report said that the horse bit his owner's jugular vein and also damaged his spine, so this was obviously a vicious attack. Still, a horse attacking and killing his owner is so unheard of that even though it happened in Poland we read about it here.

Bulls do kill people with some frequency, but when they do they're almost never trying to kill; they're challenging the person for dominance. Bulls don't kill each other when they're fighting for dominance, but because a bull is so big, and because bulls use head butting to win dominance contests, the human gets crushed against a fence. The bull doesn't understand how much bigger and stronger he is.

Even though you can handle aggressive behavior in most, *though not all,* prey animals, it's always better to keep aggressive behavior from developing in the first place. With prey animals that means good training and socialization, but not dominance training per se. I think in the old days a lot of animal handlers didn't understand the difference. They thought any kind of training was also dominance training because the trainer was in charge. That's probably where the idea of breaking a horse's spirit came from. You shouldn't break any animal's spirit, horse or dog, but a nervous prey animal like a horse or a cow doesn't need to learn obedience as a separate concept the way a dog does. A cow or a horse who's being trained just needs training, not dominating; a dog needs training, but he also needs dominating. A dog needs an alpha, or else he'll be the alpha himself. With prey animals even an aggressive, nippy kind of horse *usually* isn't much of a problem to manage.

It's *never* easy to manage an aggressive dog. The only person equipped to deal with an adult dog who bites is a professional who specializes in aggression, and even then your chances of turning the dog around are not good. Dr. Bonnie Beaver, a veterinary animal behavior specialist at Texas A&M, says that a typical case of dominance over humans gets worse, and Dr. Dodman, who treats dominance aggression, reports that only two out of three of dogs with dominance aggression end up getting a lot better, even with a formal retraining program. The other third still have problems, although most of them are safer to be around than they were. But many dogs do not improve at all. These are dangerous animals.

It's also easy for most dogs to become biters if they're allowed to hold alpha status over their owner. We don't know exactly why it's so easy to teach a dog not to bite in the first place, but so hard to teach a dog to stop biting once he's started. Why can't you turn back the developmental clock and retrain an aggressive dog the way you train a puppy?

Dr. Dodman has done research showing that in some cases the problem is the owner.[23] "Emotional" owners aren't as successful at retraining a dominant aggressive dog as "rational" owners who can stick to a retraining program. Maybe people who were too "kind-hearted" to be firm trainers and disciplinarians in the first place can't suddenly turn themselves into good trainers just because an animal behaviorist has told them that they have to do it. If they had established themselves as the alpha early in their dog's life, they would not have a biting dog. This is true for all *normal* dogs. There are some dogs who are genetically bad, dangerous dogs, the same way

174Animals in Translation

the rapist roosters were genetically bad, dangerous birds. Such dogs have to be euthanized. But if you own a normal dog, you can prevent aggression by doing enough obedience training to establish yourself as the alpha.

I think the main reason you can't train a dog back out of aggression as easily as you can train a dog into aggression is that the genie has been let out of the bottle. All dogs have a natural drive to be the alpha. Owners have to teach their dogs to think that it's impossible for a dog to dominate a human. It's not just a bad thing, it's an *impossible* thing. Once a dog has discovered that he can dominate people there's no turning back. You can't un-teach this knowledge; you can only try to teach a biting dog to inhibit his impulse to compete with his owner for dominance.

This is what happens with the big cats. In *Out of Africa* Isak Dinesen tells a story about a young pet lion named Paddy. Paddy was tame, and was nice to everyone at the ranch where they lived, although he'd never been socialized to children. Then one day someone brought a little girl to visit, and Paddy accidentally knocked her down. He didn't hurt the little girl, and he didn't do it on purpose.

But that very night Paddy went out into the pasture and killed a bunch of livestock, and from then on he had to live in a cage. He had learned that he was a lion, not a big house cat. That one moment of experiencing his power over another creature, when he knocked down the little girl, was enough to wake up his real nature.

Triggering a predator animal's aggressive nature is so dangerous that big cat handlers can use a trained lion or tiger only a few times in TV shows and movies if the scenes involve knocking down a human being. Even when a trained lion or tiger gently bumps against a human being *on command* it will soon become too dangerous to work with.

The moral for lion tamers is, Don't let this kitty ever find out that he weighs seven hundred pounds. You can arrest an animal's emotional development by not giving it a chance to figure out its strength and power, but you can't make him *un*-learn his strength and aggression once he knows it.

HANDLING FEAR AGGRESSION

Not all dogs who bite are dominant. *Shy biters* bite people because they're afraid, not because they are dominant. German shepherds who bite are usually shy biters. They are nervous animals.

Shy biters are somewhat less dangerous than dominant biters. They are dangerous mainly when the owner is around to give them courage. If a shy biter sees a stranger or a neighbor he's afraid of when he's alone, he'll usually just try to get away. If he can't get away, he will bite the stranger from behind because that's less frightening than having to meet the person's eyes. Shy dogs will avoid eye contact with everyone but their owners at all costs. That's just as well, since if you're going to get bitten by a dog it's better to get bitten on the ankle or the thigh than in the face. All in all, shy dogs are probably not as dangerous as they seem.

A *dominant* scared dog is different. Dogs who are both dominant and fearful can bite any time and place. They will bite with their owners present, or with their owners long gone. And when they bite, they can go straight for the face. Because they are dominant by nature, running away isn't an option. They have to attack. I don't think anyone knows exactly why a shy dominant dog is as potentially dangerous as he is. Is it just because he has two different reasons, fear and dominance, to bite people, which raises the odds that he will? Is it because when you mix fear and dominance together the dog's emotions are heightened and his ability to control himself is impaired?

I do know one neutered male dog who's highly dominant and fearful. He's not a shy biter, because his owners realized how dominant he was early on and did everything right, so he knows he's not the alpha.

But he's a big problem with other dogs. He'll try to attack any dog he sees on walks with his owners, and he can never be let off a leash in public or taken to a dog park. This is a dog who was well socialized to other dogs as a puppy, and yet was so dominant by nature that he still managed to get into two fights with the neighbor's dog. He won the first fight but lost the second one, and has been acting more and more terrified of other dogs ever since.

If he were a submissive dog by nature, that might not matter, because he would just avoid looking at the dog who was scaring him. But since he's dominant by nature, the instant he feels threatened by another dog he attacks—and he feels threatened all the time. Just the sight of another dog minding its own business seems to threaten him. This dog's behavior reminds me of a well-known study of anxious children versus oppositional children. (Children with *oppositional defiant disorder,* or ODD, are kids who are so angry and disobedient that their behavior disrupts their school or home life.) Both groups of children interpret ambiguous situations as being more threatening than typical kids do, but where an anxious child copes by avoiding the threat the oppositional child will become aggressive.[24] I don't think a dominant dog is the same thing as an oppositional child, but the fearful dominant dog I know seems both to exaggerate threats *and* to react aggressively to threats once he's blown them up out of all proportion in his mind.

Regardless of what makes a shy dominant biter tick, once *any* dog has begun to bite out of fear, you have an animal who is never going to be completely safe again, because no animal can be completely trained out of fear.

—

If you've never lived with a dog, by now you may be thinking the best idea for anyone who's especially safety-minded is to just stay away from any animal larger than a small cat.

But that would be the wrong conclusion. The human relationship with domestic animals goes back a long way, and people need animals in their lives. Until recently most experts believed that humans and dogs paired up together 14,000 years ago, but more recent research on dog DNA shows that humans and dogs may have been keeping company for over 100,000 years. Dogs really are man's best friend.

The reason dogs don't kill humans more often than they do isn't that owners are brilliant trainers. A lot of owners don't know the first thing about obedience training. The reason dogs don't kill humans is that in 100,000 years of evolution dogs have developed a lot of ability to inhibit aggression against humans, and humans have developed a lot of ability to *manage* dog aggression, whether they've ever read a book on obedience training or not. I think humans have probably evolved some innate ability to read dog language, or at least to learn to read it quickly.

A friend of mine told me an interesting story about this. She adopted a puppy from an animal shelter who quickly began showing signs that he was destined to be a highly dominant dog. When the puppy was only a few months old it started to growl at her seven-year-old son. A couple of weeks later the puppy bared his teeth and growled at a six-foot-four plumber who came to the house to fix the toilet.

The first time the puppy growled at her son my friend was sitting in another room and she called out to her son, "Why did Buddy growl at you?"

Her son, who had never lived with a dog in his life, said matter-of-factly, "Because I was on his chair."

He was right. Buddy had growled because he was lying comfortably on his favorite chair, which naturally was the biggest, softest chair in the house, seeing as how he was such an alpha kind of guy— *and then the boy came in and sat down on it with him!* Buddy didn't like that, and he told the boy he didn't like it in no uncertain terms.

And the boy understood. He knew exactly why his family's new dog had growled at him without having to be taught—without even having to think about it. He got the message.

Through all the years dogs have been living with humans they've developed a lot of ability to read people, to know what people are thinking and what they're likely to do. We know this from research comparing dogs to wolves. Even a wolf who has been hand-reared by human beings never acquires the ability to read people's faces the way any normal dog does. A human-reared wolf mostly doesn't look at his master's face, even when he's in a situation where he could use his master's help. Dogs *always* look at their owner's faces for information, especially if they need help.[25]

I think that as dogs were learning how to read us, we were learning how to read them. The reason dogs don't hurt people more often is that dogs and people belong together.

5. Pain and Suffering

People who love their pets usually feel like they have a pretty good idea what an animal needs to have a good life. The basic necessities of life for pets are the same as they are for us: food, safety, companionship.

That's a good start, but if that's all you know about animals you can still get into trouble. Just to give you the first example that pops into my head: anyone who's gone out and bought himself a Border collie—or who's thinking about going out and buying himself a Border collie—is missing one big item from the Border collie list, and that is a *job*. Border collies aren't built for a life of leisure, and they can get nutty if that's what you give them. Unfortunately, a lot of people don't find this out until after they've got the dog. Then they have to spend the next ten years trying to give their pet something useful to do.

It's doubly hard for ranchers, feedlot managers, and sometimes even veterinarians to know exactly what they should do to treat the animals in their care responsibly. What does a cow headed to slaughter need in order to have a happy life?

If I had my druthers humans would have evolved to be plant eaters, so we wouldn't have to kill other animals for food. But we didn't, and I don't see the human race converting to vegetarianism anytime soon. I've tried to eat vegetarian myself, and I haven't been able to manage it physically. I get the same feeling you get with hypoglycemia; I get dizzy and light-headed, and I can't think straight. My mother is exactly the same way, and a lot of people with processing problems have told me they have this reaction, too, so I've always wondered if there's a connection. If there's something different about your sensory processing, is there something different about your metabolism, too?

There could be. It's possible that a brain difference could also involve a metabolic difference, because the same genes can do different things in different parts of the body. A gene that contributed to autism might contribute to a metabolic difference, or any other kind of difference. Parents have *always* said that their autistic children have lots of physical problems, too, usually involving the gut, and mainstream researchers haven't paid a lot of attention to this.

So until someone proves otherwise I'm operating from the hypothesis that at least some people are genetically built so that they *have* to have meat to function. Even if that's not so, the fact that humans evolved as both plant and meat eaters means that the vast majority of human beings are going to continue to eat both. Humans are animals, too, and we do what our animal natures tell us to do.

That means we're going to continue to have feedlots and slaughterhouses, so the question is: what should a humane feedlot and slaughterhouse be like?

Everyone concerned with animal welfare has the basic answer to that: the animal shouldn't suffer. He should feel as little pain as possible, and he should die as quickly as possible.

But although the principle is obvious, putting it into operation isn't, because it's hard to know how much pain an animal feels. It's hard to know how much pain a person feels when you get right down to it, but at least a person can tell you in plain language that he feels horrible. An animal can't do that.

The problem isn't just that animals don't talk. Animals also *hide* their pain. In the wild any animal who's injured is likely to be finished off by a predator, so probably animals evolved a natural tendency to act as if nothing's wrong. Small, vulnerable prey animals like sheep, goats, and antelope are especially stoic, whereas predator animals can be big babies. Cats can yowl their heads off when they get hurt, and dogs scream bloody murder if you happen to step on their paws. That's probably because cats and dogs don't have to worry about getting killed and eaten, so they can make all the noise they want.

Prey animals can be incredibly uncomplaining. A few years ago my student Jennifer and I saw a bunch of bulls being castrated. The vet was using a rubber band procedure, wrapping a tight band around the bull's testicles and leaving it there for several days. That sounds horrible, but vets use it because it's less traumatic than surgery, although there are individual differences in how cattle react to it. Some bulls act perfectly normal, while others repeatedly stamp their feet. I interpret foot stamping as a sign of discomfort but not overwhelming pain.

A few bulls, though, act as if they're in agony. They lie down on the ground in strange, contorted positions and they moan—but they do this only when they're alone. When we were at the lot, one of the bulls was having a bad pain reaction, and when Jennifer walked up to his pen he jumped to his feet and greeted her as if nothing was wrong. The other bulls, who didn't seem to be especially bothered by the procedure, didn't change their behavior one way or another. When they thought they were alone—I was watching them from inside the scale house so they couldn't see me—they didn't act any different.

Sheep are the ultimate stoics. I once observed a sheep who'd just had excruciating bone surgery. I would have had no way of knowing how much pain that animal was in based on the way she was acting, and a hungry wolf would have had no reason to pick her out of a flock. An injured animal in terrible pain will actually *eat food*— something all our theories of stress tell us shouldn't happen. Physiologically, bad injuries and pain are severe forms of stress, and severe stress normally diverts bodily resources away from eating and reproduction. I warn vets about this all the time: there's no way to know how much pain an animal is in when you're right there in the room with him. Animals mask pain.

Predator animals like dogs are less likely to mask their pain, but even they do it to some degree. Pain masking may be why a lot of vets will neuter a female dog and send her home without any painkiller. Any human who's ever had abdominal surgery will tell you it's agonizingly painful, but vets say that dogs sure don't act like they're feeling anywhere near as bad as a human does. We don't know whether they're masking their pain or whether they just don't feel as much pain as we do in the first place. Either way it's a problem, because animals need some pain to keep them quiet so they can recover. If dogs do mask surgical pain it's especially dangerous, because a dog won't spend any time alone if she can help it. A lot of vets will tell you they don't like to give pain medication because they want your dog to have *enough* pain to slow it down for a while. That's not a concern you'll ever hear from a doctor who operates on humans.

A friend of mine found this out the hard way. She had a young female Lab who was used to playing with three other young dogs. You put four very young dogs together, and you've got some wild and woolly play, which is what went on every day in my friend's backyard. The Labrador had her surgery in the afternoon, then went home the same night. She was groggy and out of it, but the first thing she did when she got home was jump up on the sofa at the end of her owners' bed and from there up onto the bed. No human being five hours out of abdominal surgery will jump onto a couch, *ever.* That's something you just don't see.

So my friend and her husband gave the Lab doggie tranquilizers for a couple of days to keep her quiet, but she still played so vigorously with the other dogs that she didn't heal properly. Instead of developing a thin red scar where the incision had been made, the surgical wound kept getting wider, turning into a concave area of shiny, moist tissue.

Unfortunately, my friend didn't know what the wound was supposed to look like and didn't realize until almost too late that it wasn't healing right. She was inspecting the wound every day to see if it looked infected, and while it didn't look good to her, the incision didn't look infected, either. She was getting more and more worried, but she thought she was just being an anxious owner.

Finally she got so worried she took her dog back to the vet. He took one look at the dog's belly and told my friend that if she hadn't come in that day her dog's intestines would have been "lying on the floor" by nighttime. There was no infection, but the skin tissue was completely broken down, and there was only a thin veneer of it left holding the viscera inside. My friend was horrified. You can see why vets worry about too little pain instead of too much. That Lab could have died from a routine spaying procedure all because she wasn't showing any pain, so she didn't slow down her social life with the other dogs for even one day.

Do Animals Hurt?

The short answer is yes. Animals feel pain. So do birds, and we now have pretty good evidence that fish feel pain, too.

We know animals feel pain thanks both to behavioral observation and to some excellent research on animals' use of painkillers. Starting with behavior, dogs, cats, rats, and horses all limp after they've hurt their legs, and they'll avoid putting weight on the injured limb. That's called *pain guarding*. They limit their use of the injured body part to guard it from further injury. Chickens who've just had their beaks trimmed peck *much* less, another obvious form of pain guarding. (Ranchers trim chickens' beaks because chickens get in horrible fights and will peck each other to death. The vet trims off the sharp point so the chicken can't use it as a knife blade.)

We think insects probably don't feel pain, by the way, because an insect will continue to walk on a damaged limb.

Up until recently nobody knew whether fish felt pain or not, but two researchers in Scotland have shown that they almost certainly do. The study used electrical measurements of the brain backed up by behavioral observation. First they anesthetized some fish and applied painful stimuli like heat and mechanical pressure to their bodies while running a brain scan. They found neurons in the fish's brains that fired in a pattern very close to pain firing in a human brain. Assuming this study can be replicated, it shows that fish have at least the *sensory* component of pain, though it doesn't tell us whether the fish were actually feeling it consciously. Humans with certain kinds of brain damage can have the sensory component of pain without the "suffering" component, which I'll get to in a moment.

In the second part of the study, the researchers used behavioral observation to figure out what the fish were probably feeling. They injected either bee venom or vinegar into the fishes' lips, which would be painful for humans and other mammals, and then watched to see what the fish did. The fish acted exactly the same way mammals act when they're in pain. It took the fish an hour and a half longer to begin eating again than it did fish who'd had painless saline water injections, a classic sign of pain guarding. Their lips hurt, so they didn't want to eat. They also showed other signs of pain. They rocked their bodies, something you see zoo animals do when they hurt, and they kept rubbing their lips against the side and bottom of the tank.

These obvious behavioral changes are strong evidence that the fish were consciously experiencing pain, although the fish brain is so different from the mammalian brain that we can't say for sure. Fish don't have any neocortex at all, and most neuroscientists think you have to have a neocortex to be conscious. Still, the fact that a fish doesn't have a neocortex doesn't have to mean that a fish isn't conscious of pain, because different species can use different brain structures and systems to handle the same functions.

We have more evidence that animals feel pain from the experiments Francis C. Colpaert did on animals and pain medication in the early 1980s. He injected rats with bacteria that produce a temporary bout of arthritis we know is painful in humans, then gave them a choice between a bad-tasting liquid analgesic and a sweet, sugary-tasting liquid rats normally like. The rats chose the bad-tasting painkiller over the sugar solution, a pretty good sign they were choosing it for its painkilling properties. They definitely weren't choosing it for its taste.[1]

Once their arthritis cleared up they switched to the sugar drink, another sign they were using the painkiller to treat pain. If they'd been choosing the painkiller just because they liked it—maybe the same way humans can use painkillers as a recreational drug—they would have kept on using it after their arthritis cleared up. But they didn't. When their joints were inflamed they chose a yucky-tasting painkiller; when their joints returned to normal they stopped choosing the yucky-tasting painkiller.

Somebody needs to do Colpaert's experiment with fish. That would tell us a lot.

How *Much* Does Pain Hurt?

I think the real question isn't whether or not animals (and birds and fish) feel pain. It's pretty obvious they do.

The real question is *how much does pain hurt*? Does an animal with the same injury as a person feel *as bad as* a person does? We should be talking about degrees.

I think the answer to whether the same injury in an animal feels as bad as it does to a person is often no, for a couple of reasons. For one, even when they're alone animals usually—not always, but usually—act as if an injury or disease hurts them less than the exact same injury or disease would hurt a person. That's important.

Beyond that, a lot of what we know about the brain leads me to think animals may have a different experience of pain than people do. I remember being struck a year or so ago when I came across a study saying that chronic pain is associated with *widely spread prefrontal hyperactivity.*[2] That surprised me. Pain seems like such a basic sensation I'd just naturally thought of it as a primitive reaction all creatures have to have to protect themselves from injury. To me, pain seemed like an ancient, *lower-down* brain function. Since the frontal lobes are as high up as you can get, I wasn't expecting to read that pain is associated with high frontal lobe activity. That study made me wonder whether an animal's conscious pain may be less intense than a person's, because an animal's frontal lobes are smaller and less developed.

When I started looking into the literature on frontal lobes and pain I found out that psychiatrists have known about this connection for years. The idea that active frontal lobes mean active pain was so well established that in the 1940s and 1950s a few psychiatrists began treating cases of severe and intractable chronic pain by surgically disconnecting the patient's frontal lobes from the rest of his brain. The operation they did was called a *leucotomy,* and basically it was a less-invasive lobotomy. Where a lobotomy removed the frontal lobes completely, a leucotomy left the frontal lobes in place but cut the connections between the frontal lobes and the rest of the brain.

Both operations had a lot of horrible side effects, but the positive effect on a pain patient's suffering was almost miraculous. A couple of days after the operation patients who'd been completely disabled by pain would be up and about, doing the things they used to do. The "recoveries" were so dramatic that Antonio Egas Moniz, who invented the operation, won the Nobel Prize for his work in 1949.[3]

I put "recovery" in quotation marks because leucotomy patients

didn't exactly recover. They *acted* like they'd recovered, but whenever people asked how they felt, they'd always say the pain was still
there. What was different after surgery wasn't the pain; it was their
feelings about the pain. They didn't care about it anymore. Antonio
Damasio has a description of one of these patients in his book
Descartes' Error.[4] The first time Dr. Damasio saw him, the patient
was in such bad shape he was "crouched in profound suffering,
almost immobile, afraid of triggering further pain." Two days after
the operation the man was sitting in a chair, playing cards with
another patient. He looked completely relaxed.

When Dr. Damasio asked the patient how he was doing his
answer was, "Oh, the pains are the same, but I feel fine now, thank
you." You can read story after story exactly like that one in the literature on leucotomy and pain. After their operations, leucotomy
patients stopped *caring* about their pain. Dr. Damasio says they kept
their pain but lost their suffering.

It's impossible to imagine what it would feel like to have severe
pain but not be bothered by it, because for the rest of us severe pain
means severe suffering, period. They aren't two different things. I'm
sure that's because our frontal lobes integrate sensory pain pathways
so totally with frontal, emotional pathways of suffering that we can't
perceive any separation at all. It's a little like stereoscopic vision: if
your vision is working right you *can't* separate what your right eye is
seeing from what your left eye is seeing without closing one eye.

Even though we can't feel what the leucotomy patients were feeling, it *seems* like they were still feeling something like what we would
call pain, because they still asked for painkillers. On the other hand,
after the operation they stopped asking for really strong painkillers
like morphine. All they needed was aspirin. It's possible they were
feeling something similar to what the rest of us feel when we have
pain mild enough to ignore. Mild pain is still pain, but it doesn't
ruin your life, whereas severe pain hijacks your attention system.
That's almost the definition of severe pain, that it commands all of a
person's attention.

Another piece of evidence the leucotomy pain patients were still
feeling "real" pain, at least to some degree, is that if you suddenly
poked one of these patients with a pin they would shriek in pain.

They would actually shriek *louder* than a normal person with normal pain perception. Most researchers chalk this up not to greater pain but to lower *impulse control*. The frontal lobes censor and control outbursts of any kind, including screams of pain. Since these patients had lost their mental brakes, they screamed at a mild poke.

I think injured animals are probably somewhere in between a leucotomy patient and a normal human being. They do feel pain, sometimes intense pain, because their frontal lobes haven't been surgically separated from the rest of their brains. But they probably aren't as upset about pain as a human being would be in the same situation, because their frontal lobes aren't as big or all-powerful as a human's. That's why they don't slow down after surgery the way we do. They don't feel bad enough to slow down. I think it's possible that animals may have as much pain as people do, but less suffering.

AUTISM AND PAIN

A lot of autistic people are the same way, which is another reason I tend to think animal pain is less severe than human pain *on average*. As I've mentioned more than once, whenever I come across a difference between animals and normal people that involves the frontal lobes, I've usually found the same difference in autistic people. We have a lot in common with animals. So I'm expecting to find the same thing with pain perception.

Just like animals, quite a few autistic people—not all, but many—*act* like they feel less pain than nonautistic people. This happens so often that *insensitivity to pain* is listed on most symptom checklists for autism. It's especially shocking with little kids who are self-injurious. Some of them can slap their heads hard with their hands and not seem to feel any pain at all (other autistic children slap their heads and then cry). There've even been reports of autistic children burning their hands on hot stoves and not reacting, although fortunately that's extremely rare. Autistic children don't have such low pain sensitivity that they're in danger of injuring themselves without knowing it.

Another interesting thing: a lot of parents tell me their autistic children don't have normal sensitivity to cold, either. They can spend hours in the deep end of a freezing cold swimming pool while all the other kids just splash around for a few minutes and then go warm up on the deck. I don't know whether animals have lower cold sensitivity on the whole. Animals in northern climates do better in winter cold than people do, but they have nice fur coats to keep them warm and people don't. A wolf's coat is so thick snow doesn't melt on its body.

So I have no way of knowing how cold perception compares between animals and people with autism. Also, I want to make sure I'm not implying to parents or teachers or anybody else that autistic people are impervious to everything that comes their way. The autistic sensory system is abnormal for a person, while an animal's sensory system is normal for an animal, so I don't know where the similarities begin and end. I do know that while some things are less painful for autistic people than for typical people, other things, especially certain types of sounds, are more painful. I remember one autistic woman saying she found the sound of the ocean excruciating. (There might even be stimuli that are more dangerous for people with autism, though we don't know that. A few years ago I talked to a woman involved in autism research who said she was concerned that some autistic people might be more susceptible to heat stroke. I'd never heard that before, and she based her comment on just a couple of families, so I don't want parents to start worrying about it. I bring it up because I don't want to minimize the discomfort an autistic person could be feeling.)

I don't remember how I reacted to pain as a child, but as an adult I've been told that I'm a lot less sensitive to pain than nonautistic people. When *I* was "spayed" (I had a full hysterectomy, medically the exact same procedure as spaying a dog, that left an eight-inch scar across my stomach) I acted more like my friend's Lab than a post-surgical human being. The nurses said I didn't use anywhere near the amount of IV painkiller other patients did. Then when I went home, I took one prescription pain pill and that was it. I didn't need any more.

In the hospital I ran a little experiment on myself. When I was sure the nurses weren't around, I got out of bed and got down on all fours like a dog. The staff would have had a fit if they'd seen me. I found out that as long as I held still my pain was a lot less than it was standing up or sitting down. Crawling felt terrible, but not as bad as walking did. Still, even on all fours I didn't feel like jumping up on a sofa, so obviously I'm not as impervious to pain as a Labrador retriever. Then again, no *dog* is as impervious to pain as a Labrador retriever, either. Labs are notorious for their high pain threshold, which is one of the reasons they make such good pets for children. A little kid can jump all over them and maul them half to death and they feel nothing. (Not that I'd recommend any child being allowed to do that. It's bad manners, and with other breeds it could be dangerous.) Try stepping on any normal dog's paw and you get an ear-splitting yelp so loud that for a moment you think you've killed your pet. Step on a Lab's paw and he doesn't even blink. Labs are built for racing through bramble and brush to retrieve game, or jumping into freezing cold water to retrieve fish. Nothing fazes them.

Back to my experiment, it's possible there's something about being a four-legged creature instead of a two-legged creature that makes the pain of physical injuries less intense. But even if that turns out to be true, I expect it's going to be only part of the explanation for why animals act as if they have less pain than we do for the same injury. Eventually we'll find out that the real explanation for the difference in behavior is a difference inside the brain.

FEAR IS WORSE THAN PAIN

A lot of effort has been put into creating humane slaughter systems so the animal doesn't suffer. That part was easy, relatively speaking. If all you had to do to eliminate suffering was to make sure the animal died instantly, today almost all of our slaughterhouses would have to be considered humane.

But eliminating pain isn't enough. We have to think about animals' emotional lives, not just their physical lives. We're responsible for slaughterhouse animals; they wouldn't even exist if it weren't for us. So we have to do more than just take away physical pain.

The single worst thing you can do to an animal emotionally is to make it feel afraid. Fear is so bad for animals I think it's worse than pain. I always get surprised looks when I say this. If you gave most

people a choice between intense pain and intense fear, they'd probably pick fear.

I think that's because humans have a lot more power to control
fear than animals do. My guess is that animals and normal humans
are opposites when it comes to fear and pain, and for roughly the
same reason: different levels of frontal lobe functioning. This idea
first popped out at me when I read two studies back-to-back on the
frontal lobes in pain and in fear.[5] What struck me was that while an
active prefrontal cortex was associated with increased pain, it was
also associated with reduced fear (though not with reduced anxiety).
Pain and fear, at least in these studies, were opposites.[6]

The story isn't that simple, of course, but it's close enough that,
until we learn more, I believe animals have lower pain and higher
fear than people do. My other reason for believing this at least provi-
sionally is that it's the same with autistic people. As a general rule,
we have lower pain, higher fear, and lower frontal lobe control of
the rest of our brain than nonautistic people. Those three things go
together. (I'm not saying that autistic people have no pain at all and
don't need painkillers. I don't want to give that impression.)

You almost have to work with animals to see what a terrible emotion fear is for them. From the outside, fear seems much more punishing than pain. Even an animal who's completely alone and giving
full expression to severe pain acts less incapacitated than an animal
who's scared half out of his wits. Animals in terrible pain can still
function; they can function so well they can act as if nothing in the
world is wrong. An animal in a state of panic can't function at all.

I also think intense fear is an easier state for animals to get into
than it is for normal human beings—a lot easier. Animals feel intense
fear when they're threatened in any way, regardless of whether
they're predators or prey.

While all animals can be overwhelmed by terror, prey animals like
cows, deer, horses, and rabbits spend a lot more time being scared
than predators do. You've heard the expression "like a deer caught
in the headlights"—that pretty much sums up the prey animal's psy-
che. They are very nervous animals, because the only way a prey animal can survive in the wild is to run. Since a prey animal has to start

running *before* the lion does, that means it has to be hyper-alert all
the time, keeping a watch out for danger.

You have to be gentle when you're working with prey animals.
I've seen so many animals ruined by owners who traumatized them
through rough or ignorant handling. The whole idea of *breaking* a
horse is a perfect example. If you break a horse, he's *broken*. He's
traumatized for life and usually no use to anyone after that, including himself a lot of times. Just like the horses at my school.

That's another thing autistic people have in common with animals: we have long memories, especially for fear. Clara Barton had a
famous saying: "I distinctly remember forgetting that." No autistic
person would have come up with that. We *can't* forget bad stuff on
purpose, and neither can animals.

I'm sure that's why I relate to prey animals like cattle as strongly
as I do: because my emotional makeup is similar. Fear is a horrible
problem for people with autism—fear *and* anxiety. Fear is usually
defined as a response to external threats, while anxiety is a response
to *internal threats*. If you step on a snake you feel fear; if you *think*
about stepping on a snake you feel anxiety.

It's not clear whether the brain system underlying fear and anxiety
is the same. I think most researchers have assumed that it is, but
recent research by Ned Kalin, a psychiatrist at the University of Wisconsin, Madison, has found a difference between our "initial
responses to fearful stimuli" and an "anxious temperament."[7] The
amygdala handles fearful stimuli, but the prefrontal cortex is responsible for an anxious temperament. When the amygdala is damaged
the anxious temperament doesn't go away.

Based on my own observation of animals and of myself, I think
nature created at least two different emotional systems to handle
threats: fight-or-flight fear and the orienting response I talked about
in Chapter 2. The fight-or-flight corresponds to fear, and I wonder
whether the orienting response might correspond to anxiety or to
the anxious temperament.

I say that because if I'm God and I'm designing an animal I don't
want to give him *only* a fight-or-flight system. I want to give him
vigilance as well, because I want him to keep a lookout. I need two

different systems because if he just chronically flees every potential threat, he's going to use up his energy reserves. The reason I think vigilance may be linked to anxiety is that anxious people are always on guard, always watching for trouble.

I don't know what the research will show, but I do know that antidepressants have separated my orienting response from my fear response. One reason I say this is that I take antidepressants, and they've gotten rid of my fear but not my orienting response, which makes me think the two responses are based in separate systems in my brain. If you take something like the high-pitched back-up alarm on a garbage truck, before I took medication I would almost have a panic attack hearing that sound. On medication I don't panic, but the beeping turns on my orienting response and I can't turn it off. If I'm trying to sleep and I hear a back-up beep I can't *not* orient to it; I have to pay attention. There's no way I can fall asleep. It's as if the medication split my system apart chemically. Intense fear got turned off; orienting and hyper-vigilance stayed on.

Autistic people have so much natural fear and anxiety—I'm almost comfortable saying it's universal—that when they're young they can be like little wild animals. For years people thought autistic children were unteachable because they were so uncontrollable, and a lot of people think that the *feral children* we've heard about over the years—children said to have been raised by wolves—were actually autistic. No one would call an autistic child feral today, but the word is a pretty accurate description of the way a lot of these children—not all, but quite a lot—*appear* to normal people who've never dealt with them before.

Autistic children seem "wild" for a lot of different reasons, not all of which relate to animals. A huge problem for autistic children, though not for animals, is scrambled sensory processing. The world isn't coming in right. So young autistic children end up looking wild for the same reason Helen Keller looked wild: parents and teachers can't get through to them. In some ways it's almost like they have to raise themselves. A lot of them do a good job of it, because over the years they seem to start piecing things together. One mother told me she felt as if her son had to "learn to see," and I bet there's a lot of truth to that.

But one of the biggest reasons autistic children (and more than a few autistic adults) seem so "untamable" is that they're terrified of so many things. It can take years for an autistic child to lose his fear of the most ordinary events, like getting a haircut or going to the dentist, if he ever does. There are plenty of autistic adults who have to be given general anesthesia to have their teeth worked on. They've never gotten over their terror.

This is what we have in common with animals. Our fear system is "turned on" in a way a normal person's is not. It's fear gone wild. In my own case, overwhelming anxiety hit at puberty. From age eleven to age thirty-three, when I discovered antidepressant medication, I felt exactly the way you feel when you're about to defend your dissertation, only I felt that way all day long, every single day. I was in a constant, daily state of emergency. It was horrible. If I hadn't gone on medication I couldn't have had a life at all. I certainly wouldn't have been able to have a career.

FREEDOM FROM FEAR

It seems likely that animals and autistic people both have *hyper-fear* systems in large part because their frontal lobes are less powerful compared to the frontal lobes in typical folks. The prefrontal cortex gives humans some freedom of action in life, including *some* freedom from fear. As a rule, normal people have more power to suppress fear, and to make decisions in the face of fear, than animals or (most) autistic people.

The frontal lobes fight fear in two ways. First, the frontal lobes are the brakes. The frontal lobes tamp down the amygdala, a tiny, evolutionarily ancient structure in the middle of the brain that produces fear. The amygdala tells the pituitary to pump out stress hormones such as cortisol; the prefrontal cortex tells the pituitary to slow down. I don't know for a fact that an animal's or autistic person's frontal lobe braking system is weaker than a nonautistic person's, but my guess is that it is. We could certainly discover that different species have different levels of frontal control over fear, too.

Even if we find out animal frontal lobes do just as good a job of suppressing stress hormones as the human brain, the frontal lobes have a second means of combating fear that we know almost to a certainty is different in animals and in typical humans, and that is language. Nonautistic people use language to talk themselves out of fear.

There's probably more to it than that. I have come to believe, from my own experience and from published research, that *mental images* are far more closely connected to fear and panic than words. Ruth Lanius, an assistant professor psychiatry at the University of Western Ontario, did a brain scan of people suffering from *post-traumatic stress disorder,* or PTSD.[8] She scanned the brains of eleven people with PTSD as a result of sexual abuse, assault, or car crashes and thirteen people who had suffered the same experiences without developing PTSD. The main difference she found between the two groups was that one group remembered their trauma visually and the other remembered it verbally, as a *verbal narrative.* Their scans backed this up. When people with PTSD remembered the trauma, visual areas of their brains lit up (along with other areas), and when people without PTSD remembered their traumas, verbal areas lit up.

Somehow words are associated with lower fear. This is one of the meanings of the saying "A picture is worth a thousand words." A picture of a scary thing is a lot more frightening than a verbal description of a scary thing. By the same token, a visual memory of a scary thing is more frightening than a verbal memory. No one knows why or how words are less frightening, or how this works in the brain. But I think that when it comes to managing their fear, animals and autistic people are at a big disadvantage because they have to rely on pictures.

———

I don't know whether it's easier to traumatize an animal than a human being overall. I think it probably is. I do know that once an animal has become traumatized it's impossible to un-traumatize him. Animals never unlearn a bad fear.

There's no reasoning with an animal who's been scared half out of his wits. Here's a classic example. When she was little, a friend of mine had a collie who became deathly afraid of the basement.

Apparently the dog had gotten really sick when she was a puppy, and my friend's parents put her in the basement so she wouldn't mess up the house. Afterward the dog associated the basement with being horribly sick.

That dog *never* got over her fear of the basement, and she never set foot downstairs again for as long as she lived. It was sad, because the dad had his office in the basement, and the dog wouldn't spend any time with him at all. My friend remembers her dad standing at the bottom of the stairs, calling "Lassie, Lassie. Here, Lassie," in his softest voice. (They named the dog after the Lassie in the TV show.) Lassie would stand at the top of the stairs, staring at the dad, wagging her tail frantically, even whimpering and crying because she wanted to go down to him so badly. But she would not move. You could put a big thick juicy piece of raw steak halfway down the stairs—nothing doing. She wouldn't budge. And if someone tried to pick Lassie up and carry her down the stairs she'd get violent. This was a *collie*. This dog was so terrified of the basement she was fighting for her life.

People with severe cases of PTSD don't get over it, either, but people with milder traumas have a lot of leeway in dealing with their fears. My friend with the collie developed a mild case of PTSD herself after a car crash about six years ago. She was having semi-flashbacks while she was driving, and she felt a huge amount of tension and fear whenever she had to drive on the freeway. I say semiflashbacks because she didn't feel as if she was reliving the accident; it was more that she kept vividly remembering the accident anytime she had to drive anywhere, and sometimes even when she wasn't anywhere near a car. Her memories were all visual, just like the people in Dr. Lanius's experiment.

It took her a good two or three years, but today she's basically over it. Getting into a car doesn't automatically trigger memories of the accident the way it used to, and most of the time she takes driving for granted the way she did before the accident. An animal can't do this. No animal goes back to acting nonchalant about a person, place, or situation once he's been scared half out of his wits. It just doesn't happen.

FEARLESS GUPPIES

I don't know where autistic people fall on the trauma spectrum, although I believe animal fear is more adaptive than autistic fear, on the whole. Autistic people have way too much fear, while in most circumstances animals have just enough.

I say "enough fear" because fear has a purpose, and an animal or a person *without* fear has a disability. The purpose of fear is to keep us alive. It does an excellent job of this, judging by what happens when you're low-fear. Randolph M. Nesse and George C. Williams describe a terrific study on fear and survival in their book *Why We Get Sick*.[9] Some researcher put a bunch of guppies in with a piranha in a fish tank. Some of the guppies were highly fearful, some were moderately fearful, and some were practically fearless. The fearless guppies were the ones who would stare straight at the piranha.

The fearless guppies got eaten first. If you're a guppy and you're not afraid to swim out in the open and stare straight at a piranha, you're not going to live long. Next to go were the medium-fear guppies, who didn't stare straight at the piranha but didn't do everything in their power to get away, either. They got eaten next.

The fearful guppies lived the longest. They got eaten, too, but not until everyone else was eaten up. Fear kept them alive longer.

It's pretty obvious how fear would help keep you alive when you're a guppy swimming with a piranha. But fear is also a survival mechanism when you're a piranha swimming with another piranha. Researchers found this out doing genetic *knockout* work with mice. A knockout mouse is a mouse whose genes have been manipulated to eliminate, or knock out, just one of them. You can eliminate both copies of the gene, or just one copy. Once the gene is knocked out the researcher studies the mouse to see what's different about it, if anything.

The link between fear and survival popped out in the middle of a knockout study on learning. Six months into the study there were some strange things going on in the mouse colony. The researchers would come to the lab first thing in the morning and find dead mice in the cages. Their backs were broken and there was blood everywhere. They had obviously fought to the death, which is very unusual for mice. Mice normally either avoid fights, or end a fight before either mouse dies.[10]

The researchers discovered that they hadn't just knocked out some aspect of learning; they'd also knocked out fear. A normal mouse, with a normal amount of fear, does not fight to the death. He fights until he's beaten, or sees he's going to lose, and then he yields. Fear keeps him alive. The knockout mice were almost fearless, and they fought to the death.

The researchers discovered some other interesting things about their mice. A normal mouse will fight an intruder on his territory. Lab experiments on this are clear: if you put a strange mouse inside another mouse's cage, the *resident mouse* will attack. That's called *defensive aggression,* because the resident mouse is fighting to defend his home. A normal mouse who's forced to *become* an intruder—a mouse who suddenly gets put inside another mouse's cage—*won't* fight. It will either run away or stand up in a defensive position to protect itself.

Mice with just one copy of the knockout gene were different. (Mice with both copies of the gene knocked out were messed up in so many ways that I'm not going to talk about them here. They had huge problems in lots of realms, which makes it harder to say anything specific about fear.) These mice showed normal defensive aggression if a stranger mouse was put into their cage. They fought to protect their home territory. But they also fought when *they* were the intruder. They'd get dropped into a strange cage already "owned" by a strange mouse, and instead of trying to run, they'd fight. Not only did they attack the resident mouse, but after an initial skirmish they'd approach the resident mouse *again* and start a whole new fight. This is something you'd never see in a normal mouse. A normal mouse finding himself on another mouse's territory would be too frightened to fight.

The experimenters knew the problem was reduced fear because they ran a bunch of other tests showing that these mice were less fearful in all kinds of situations. One example: the knockout mice didn't freeze up as much as normal mice do when put in a cage where they'd been shocked. The experimenters also ran experiments showing that the mice remembered the fact that they'd been shocked just fine. To test whether or not mice remember the place where they've been shocked, experimenters use a *shuttle box*, which is a cage with a partition in the middle that the mouse can jump over. The mouse only gets shocked on one side of the partition but not the other. All mice quickly learn which side is the bad side, and they stay on the good side. They remember which side is good and which side is bad.

The knockout mice remembered the shocks and they remembered that the shocks hurt. They just didn't care. And they didn't care because they didn't have the right amount of fear. If these mice had been living in the wild they would have had very short lives.

STAYING ALIVE

That's the point of emotions: survival. Normal emotions are essential to staying alive and well. Emotion is so important that if you had to choose between having an intact emotion system in the brain and having an intact cognitive system, the right choice would be emotion. Emotions are so important that, as Jaak Panksepp says, "there are good reasons to believe that the cognitive apparatus would collapse if our underlying emotional value systems were destroyed."[11]

To most people, this doesn't make sense. We humans tend to think of emotions as dangerous forces that need to be strictly controlled by reason and logic. But that's not how the brain works. In the brain logic and reason are never separate from emotion. Even nonsense syllables have an emotional charge, either positive or negative. *Nothing is neutral.* That's what you have to remember.

I want to stick with this idea for a little bit longer, because it's important to understanding what fear means to an animal. The reason most people think logic is more important than feeling is that we aren't usually aware of the connection between the two. A lot of people's emotional life is unconscious a lot of the time, especially when you're calmly thinking something through. You *feel like* you're just using logic, but you're actually using logic guided by emotion. You just aren't aware of the emotion. Not only that, but sometimes when you *are* aware of your emotions, because you're passionate about an issue or a person, you make bad decisions and you blame the emotion. And, of course, most of us definitely think *other* people's emotional decisions are dumb!

We're half right about all of this. A lot of obviously emotional decisions probably *are* dumb a lot of the time. But the problem isn't the fact that emotion was involved. Everyone uses emotion to make decisions. People with brain damage to their emotional systems have a horrible time making any decision at all, and when they do make a decision it's usually bad. The problem isn't the emotion; the problem is that the emotions they're using are dumb.

I recommend *Descartes' Error* to anyone who's interested in emotions, intuition, and decision making. Dr. Damasio has done a huge amount of work with frontal lobe patients who lost the ability to have what we call a *gut feeling*. Even though these patients still had completely normal IQs and logical reasoning abilities, they couldn't function as normal adults. They needed other adults to take care of them, and after Dr. Damasio testified for one of them in court he qualified for permanent disability payments.

The really interesting thing is *why* these people can't function. On paper, it seems like they ought to be able to manage their lives just fine. They pass most or even all of the standard neuropsychological tests. Elliot, Dr. Damasio's patient, had a high IQ and tested well on "perceptual ability, past memory, short-term memory, new learning, language, and the ability to do arithmetic."[12] His attention was good, and so was his working memory.

Working memory, by the way, is the part of your memory that *performs work*. When you hold a phone number in mind while you dial you're using working memory. Or, if you're a researcher or a writer, working memory holds two different ideas in mind while you're trying to figure out how they're related. Working memory also searches your brain for any other ideas that might be related to the first two. In other words, working memory is in charge both of *finding* things in long-term memory and of *holding* those items in conscious memory so you can use them once they're found. If you've got working memory deficits—which I do—it's a problem.

Elliot's working memory was fine. All of his cognitive abilities tested out fine, too; on paper there wasn't anything wrong with him.

It took Dr. Damasio a long time to figure out what it was Elliot *couldn't* do, and what he found relates directly to animals and their emotions. What Elliot couldn't do was have a proper emotional response to life. Dr. Damasio writes, "I never saw a tinge of emotion in my many hours of conversation with him: no sadness, no impatience, no frustration with my incessant and repetitious questioning. . . . [In the rest of his life] he tended not to display anger, and on the rare occasions when he did, the outburst was swift; in no time he would be his usual new self, calm and without grudges."[13]

Elliot hadn't just lost the big emotions like fear and anger, either. He'd lost his *visceral* emotions, the kind of thing you feel when you look at a photograph of a terrible accident or a wounded animal—or, on the positive side, a happy child or a sunset. He was mostly an emotional blank.

Why was that such a huge problem? *Because people and animals use their emotions to predict the future* and make decisions about what to do. That's what Elliot couldn't do after his brain damage: he couldn't predict the future, so he couldn't decide what to do about the future. He'd get stuck in endless deliberations instead. One time when Dr. Damasio asked him what day he wanted to come to the office next week, Elliot pulled out his date book and spent a full half-hour going through all the pros and cons of each one of the two days Dr. Damasio had suggested. He went on and on and on, spelling out all the possible consequences of either choice and never reaching a conclusion. Finally Dr. Damasio just picked one of the days. Without visceral emotion, Elliot couldn't automatically predict which day would be better and which day would be worse; he also couldn't tell whether the two days would be equally good or equally bad. He couldn't decide about the future.

If he *did* manage to take an action, it was almost always the wrong one. His judgment was shot.

But he sure did great on all those tests. Eventually one of Dr. Damasio's graduate students developed a test that picked up the difference between Elliot and people whose brains were normal. They called it the Gambling Test. In the test, the subject, who is called the Player, starts out with $2,000 in play money and four decks of cards to draw from. All he knows about the game is that every card he turns over will win him money, but a few of the cards will also require him to pay a "fine" to the experimenters, so he takes a loss on those draws. The goal is to try not to lose any of the loaned money and to win as much extra money as possible.

What the player doesn't know is that decks A and B give you really high wins but also really high losses. Decks C and D give you lower wins and lower losses. If you could sit down and do the math you'd find out that in the end you come out ahead drawing from Decks C and D. But that's not allowed, because the Gambling Test is supposed to be like life: it's uncertain. You don't *know* what's going to happen, so you don't *know*, for sure, what to do. You have to rely on intuition; you have to develop a feeling for which decks are the good ones.

That's what emotions do. Emotions let you develop hunches. They give you a feeling—and it really is a *feeling*—for what's going to happen in the future so you can make the right decision about what to do.

Elliot flunked the test. He started out like everyone else, picking cards from Decks A and B because the payoff was so high. But he didn't change to Decks C and D when he noticed his money dwindling down to nothing. People with normal brains, and even people with other kinds of brain damage (including language disorders!) start to get a bad feeling about Decks A and B pretty quickly. Once they have that bad feeling they switch to Decks C and D. But Elliot never switched. Although he understood perfectly well that he was losing his shirt, he never got a bad feeling about A and B, so he never switched to C and D. He just kept on picking cards from A and B and going deeper into debt.

USING EMOTIONS TO PREDICT THE FUTURE

A healthy animal is the exact opposite of an emotional blank, and he makes sound, emotion-based decisions all the time. He has to; otherwise he'd be dead. The single most important thing emotions do for an animal is to allow him to predict the future. We didn't always know that, but thanks to research we do now.

Animal behaviorists have learned that emotions work a lot like hunger. It's easy to see that the whole point of hunger is to keep you alive and functioning. Hunger makes you get up off your comfy seat on the sofa, or up off your comfy seat on a rock inside your cave, and go find something to eat. But what most people don't know is that hunger isn't just a *motivator* of action, it's also a *predictor* of the future. Your body doesn't wait until the last possible moment to get hungry. Instead you get hungry long before you're in danger of running out of the energy you need to keep on finding and consuming food. *Hunger is an early warning system.*

Nature is filled with systems just like the hunger system, and that includes our emotional system. Emotions don't just give you motivation; they give you *information*—information about the future and what you need to do about it.

The way our bodies work reminds me of a question productivity consultants ask companies about when they deal with problems. Do they deal with a problem when it shows up, or when it blows up?[14] The right answer is "when it shows up." Companies that wait until the last minute to handle a problem end up handling a much bigger problem than they would have if they'd jumped on it as soon as they knew about it.

It's the same way with nature, only nature goes management consultants one further: Mother Nature tries to keep us out of trouble in the first place. This isn't speculation. We know that emotions work by letting animals predict the future thanks to research into fear and the sense of smell in rats.[15] All mammals have two systems for smell: a close-up system (called the *accessory olfactory system,* or AOS) and a distant system (the *main olfactory system,* or MOS). The close-up system is extremely close-up; an animal almost has to be touching an object to smell it using the AOS.

Although it's not a mammal, the snake is a good example of the AOS. Snakes smell the air by flicking their tongues in and out. When they do that they are actually catching air molecules on their tongues and moving them to the roofs of their mouths, where the AOS is located.[16]

When it comes to picking up a predator's scent, the close-up system lets a rat smell a cat that's sitting no more than a foot or two away. The distant system lets rats smell a cat way off in the distance.

So naturally everyone assumed that rats—and any animal who's vulnerable to being attacked and eaten—would use their distant smell system to stay out of danger. It just stood to reason that if you're a rat and you don't react to a cat until you're face-to-face with it, it's too late. You're lunch.

But it turns out that's not the way things work at all. The distant system *isn't* connected to fear centers in the rat's brain, and the smell of a predator in the distance does *not* motivate a rat to flee. The distant smell system doesn't affect a rat emotionally or behaviorally at all.

It's the close-up system that's connected to the fear centers in the rat's brain, and it's the close-up system that activates survival behaviors like freezing in place or fleeing. It's the close-up system that keeps rats alive. We know this from experiments comparing rats whose close-up system has been disconnected from the rest of the brain to rats whose long-distance system has been disconnected. (This is done by snipping the fibers connecting the two inside the brain.) Only rats who have an intact close-up smell system act scared when they smell a predator. The instant they smell cat they freeze and start dropping more pellets of poop, classic signs of fear. The rats whose brains are getting input only from long-distance smell don't react at all. They feel nothing emotionally.

Researchers were stunned to get this result. It was completely counterintuitive, because why would nature want a rat to wait to get scared until he's standing face-to-face with a cat?

The answer is nature wouldn't want that, and that's not what nature did. What nature did by linking close-up smell to fear was to give the rat *the ability to predict the future.*

Here's how it works. In the wild, rats get scared when they wander into a place where a predator has been in the past. There's no cat there now (or let's hope not), but there's plenty of cat smell, and the rat is right on top of it when his close-smell system picks up the scent. Since most predators are territorial, where a cat has been in the past is an excellent indication of where it's going to be in the future. So the rat's close-up "scary smell" system lets it predict where any cats in the area are going to be and then get out of the way before they get there. It's an early warning system. Animal emotions help animals stay out of trouble in the first place, which is a very good idea if you're a rat. It's probably a good idea if you're a dog or a cat, too. Cats might want to stay away from major dog spots, and dogs who've lost fights might want to stay away from spots the victor dog is going to be visiting soon.

It seems like Mother Nature thinks an ounce of prevention is worth a pound of cure. And emotions are essential to prevention. A healthy fear system keeps animals, and people, alive by allowing them to predict the future.

When you think about emotions as a prediction system, it stands to reason that close-up smell would be wired to fear. But it's still not obvious why nature would wire up a rat's brain so that it *doesn't* feel fear when it smells a real live cat off in the distance. Shouldn't a rat who *knows* there's a cat in the detectable distance be motivated to put even more distance between itself and death-by-cat?

I don't think so. Fear is such an overwhelming emotion for an animal that evolution probably selected for brain systems that keep it under control. Propagating a species takes more than just not getting eaten. All creatures need to eat, sleep, mate, have babies, and feed and protect the young until they're big enough to fend for themselves. To do all that, a rat has to have some time off from fear. If rats froze in place every time they smelled a cat in the distance they could be frozen around the clock, depending on the neighborhood where they live.

My explanation is ad hoc, of course. You can't know why one thing evolved and another thing didn't, and it's a mistake to assume that everything we see in nature serves a purpose. Evolution can be random, and some things are probably just the side effects of other characteristics that did give animals an edge when it came to survival. But I think wiring close-up smell to fear probably did confer an evolutionary advantage. Until someone else comes up with a better idea, it makes sense to me.

The same basic principle (close-up = fear; distant = calm) probably applies to other senses as well. Take vision, for instance. People are always struck by how nonchalant prey animals are when they see a predator who can't get to them—not just nonchalant but sometimes downright provocative. A friend of mine once watched a squirrel in a tree tease a cat way down on the ground for a full half hour. The squirrel would creep down the trunk, getting closer and closer to the cat, looking it straight in the eye, until finally the cat sprang. Then the squirrel would scamper back up to safety and the cat would have to drop back down to the ground, because the trunk was too long for the cat to make it all the way up to where the branches began. There's no way for me to know what was in that squirrel's brain, but to my friend it sure looked like the squirrel was deliberately taunting the cat. He definitely wasn't frightened, because a frightened squirrel, just like a frightened rat, displays very specific behaviors like freezing in place. This was not a frightened squirrel.

He was definitely using his vision (he was probably smelling the cat, too), because he was staring at the cat intently. So obviously the sight of a predator out of the range of danger does not activate a squirrel's fear system. I suspect that if you surgically removed his close-up smell system and put him eyeball-to-eyeball with a cat, he'd panic. Distant predators don't fire up fear; close predators—or close *signs* of predators, like smell—do.

You see the same thing with dogs. Dogs know when other dogs are on a leash. Another friend of mine lives with a young male dog named Jazzie who's part Rottweiler. Jazzie is an extremely dominant dog, so he's always trying to get in fights. My friend's husband says Jazzie takes offense at any dog who's not "minding his own business," which means a dog foolish enough to look Jazzie in the eye. According to Jazzie, a dog who happens to enter his body space is supposed to bow his head and avert his eyes. Maybe a cat can look at a king, as the saying goes, but a dog definitely cannot look at Jazzie. He's going to get chomped if he does.

Jazzie lives next door to an unneutered golden retriever named Max whom he's tangled with a couple of times. For a while after that everything was fine because Max acknowledged Jazzie as the leader. Whenever Max got within a certain distance of Jazzie he would avert his eyes and then, if the distance between them got even smaller, drop to the ground. Both dogs seemed to know how close it was okay for the two of them to be without Max having to look away or drop to the ground.

But if Jazzie happened to be on a leash, forget it. Max would drop all his submissive behaviors and carry on like Jazzie was no more threat than a flea. Max would also act outrageous whenever Jazzie was behind the sliding glass doors looking at him. My friend said it was hilarious watching the two of them. Max would look straight at Jazzie—just like that squirrel—then wander nonchalantly around the deck, peeing all over the place.

It's the same story with deer, who are some of the most timid animals on the planet. Jazzie's house has an invisible fence and the deer know exactly where the electronic perimeter is. They'll calmly stand outside the boundary munching grass. Every once in a while they'll give Jazzie a direct stare, a challenging behavior no prey animal would ever do to a dog close enough to strike. Those deer know Jazzie can't get to them, so they aren't afraid. Distant-sense sensory systems do not activate fear.

The total disconnect between distance sensing and fear is really striking in the wild. A herd of antelope won't show the tiniest concern about a pride of lions sunning themselves not too far away. When you observe these animals you see that prey animals are very aware of whether or not a predator is stalking them. They know what stalking behavior looks like, and if they don't see stalking behavior they don't worry.

So we have a lot of evidence that animals are put together in such a fashion that they have a good chance of not getting frightened any more often than they have to. Nature seems to have tried to wire animals and people to have *useful* emotions, useful meaning emotions that keep us alive long enough to reproduce. Emotions keep us alive by letting us make good predictions about the future, and good predictions let us make good decisions about what comes next.

How Do Animals Know What's Scary?

There's a fair amount of research showing that certain basic fears are built into animals and people. The visual cliff experiments I described in Chapter 2, showing very young children and animals refusing to crawl or walk over what looks to them like a cliff, are an example of an innate, inborn fear. No one has to teach young humans or animals to fear heights. They already know.

More recently, Jaak Panksepp found that laboratory-reared rats who've never seen or smelled a cat stop playing the instant you put a tuft of cat hair in their play space. Since frightened animals don't play, that's a good indication that those rats are afraid. "The animals moved furtively," Dr. Panksepp says in *Affective Neuroscience*, "cautiously sniffing the fur and other parts of their environment. They seemed to sense that something was seriously amiss."[17]

This experience got Dr. Panksepp to thinking about how many research laboratories might be messing up their results due to researchers' coming to work smelling like their pet cats. The Pet Food Institute says there were 75 million pet cats living in the United States in 2002. That's a lot of cats. Since a huge amount of what we know about the psychology of learning and behavior comes from lab rats, you have to wonder how much of that knowledge came from terrified rats. This is an extremely important question, because learning done in a state of fear is different from learning done in a state of calm. I'll get to how it's different shortly.

Dr. Panksepp didn't have a pet cat, but he did have a dog, a Norwegian elkhound named Ginny. He realized he had to find out whether his own research was being affected by the fact that he was coming to work every day smelling like elkhound. So he covered his rats' play space with a massive amount of Ginny's hair, and—nothing happened. The rats kept right on frolicking and playing. Dr. Panksepp thinks this is evidence that ancient rats weren't hunted too much by ancient dogs.[18]

UNIVERSAL FEARS

We know what most of the almost certainly innate fears are. All children under the age of two are afraid of sudden sounds, pain, strange new objects, and losing physical support.[19] After age two, children lose these fears. That's decent evidence that these fears are innate. Every child has them at the same age, then every child loses them at the same age.

Older children and adults also have a set of universal fears that may or may not be innate: sudden sounds, a stranger walking toward you with an angry look on his face, snakes, spiders, dark places, and high places. Animals have whole sets of similar fears. Most mammals don't like snakes, and all animals are frightened by sudden sounds. Animals don't like anything sudden at all.

Other animal fears are more specific to each species. Mice and rats, for instance, don't like well-lit open spaces. If you plop a lab rat down in the middle of an open room in broad daylight he'll freeze and defecate. That makes sense for a small prey animal like a rat whose best bet for not getting killed is to stay out of reach and out of sight. All those old Tom and Jerry cartoons are ethologically correct: mice like mice holes. Small prey animals are happiest in small, dark places where larger predators can't get to them.

Big prey animals like cows and horses, on the other hand, are fine with wide-open spaces. They'd have to be or they couldn't get enough food to eat. If you're a thousand-pound animal trying to live on grass, you need a lot of grazing space. To stay safe, herd animals like horses and cows create their own "small space" by clustering together in groups. You'll always find the dominant animals standing in the middle of the herd where it's safest, too. That way they've got a lot of animal shields standing between them and whatever predator comes along.

Predator animals like wolves seem to be perfectly happy out in the open, but even they like to nap and sleep together inside a small den, where other predators can't get at them. In short, all animals, predator or prey, have natural-seeming fears of the natural dangers their worlds present.

IT'S EASIER TO LEARN SOME FEARS THAN OTHERS

But the story doesn't end there, because animals (and people) also have a number of fears that fall somewhere between innate and learned. These are fears that are extremely easy to pick up, like snake phobias in people. Snake phobias are common, and no snake has ever bitten most of the people who have them. Some people with snake phobias may never even have seen a snake outside a photograph. And yet they're terrified by the very thought of a snake.

That wouldn't necessarily seem like evidence that snake phobias are semi-innate if it weren't for the fact that people *don't* easily develop phobias to all kinds of things that are much more dangerous nowadays, like automobiles or electrical outlets. I'm not even sure a person *can* develop a car phobia per se. People who've been in bad accidents can and do develop post-traumatic stress syndrome, but they don't feel fear just looking at a photograph of a car, as people with snake phobias do looking at a picture of a snake. They're terrified of *riding* in a car, but the fear doesn't spread any further.

Animals show the same bias toward certain fears and against others. Psychologist Susan Mineka's experiments with monkeys and snakes at Northwestern University are probably the most important evidence we have of this. She started with the fact that monkeys living in the wild are terrified of snakes, while monkeys raised in labs are not.[20] Show a live snake to a bunch of wild-reared monkeys and they explode. They make faces, flap their ears, grip the bars of their cages, and their hair stands on end (piloerection). Wild-reared monkeys refuse to even look at the snakes; that's how aversive the presence of a snake is to a wild-reared monkey.

But show the same snake to a monkey who grew up in the lab and nothing happens. He's not worried. So obviously monkeys don't come into the world already knowing snakes are bad. Somebody has to teach them.

What Dr. Mineka showed is that it's super-easy to teach a lab monkey to be just as terrified of snakes as any monkey living out in the wild. When Dr. Mineka exposed her fearless monkeys to wild-reared monkeys acting afraid of snakes, the lab monkeys instantly got scared themselves, and they stayed scared. All they had to do was watch one snake-scared monkey, and they were snake-scared for life themselves. It took only a few minutes. Moreover, the lab-reared monkeys learned the same level of fear the *demonstrator monkeys* showed. If the demonstrator monkey was scared but not panicked, the *observer monkey* learned to be scared but not panicked, too. If the demonstrator monkey was terrified, the observer monkey learned to be terrified.

And, after learning snake fear through observation, the lab-reared monkey was just as good a fear model for other lab-reared monkeys as the wild-reared monkey had been for him.

Dr. Mineka also showed that it's impossible to teach a monkey to be afraid of a flower using the same technique. She showed her lab monkeys videotapes of a flower followed by a shot of a monkey acting terrified, making it look like the monkey on the tape was terrified of the flower. That tape had no effect. Watching a video of a monkey acting afraid of a snake scared the lab monkeys to death; watching a video of a monkey acting afraid of a flower didn't faze them.

Most researchers have concluded that the fear of snakes is *semi-*innate. Monkeys aren't born fearing snakes, but they *are* born ready to fear snakes at the first hint of trouble. Animal behaviorists call snakes a *prepared stimulus,* meaning that monkeys have been prepared by evolution easily to acquire a fear of snakes.

Dr. Mineka also found she could protect an animal from developing a fear the same way. If she first exposed a lab-reared monkey to another lab-reared monkey *not* acting afraid of a snake, that gave him "immunity." After that, if he saw a wild-reared monkey acting scared of the snake, he did *not* develop snake fear himself. He held on to his first lesson.

LEARNING BY WATCHING

This is called *observational learning.* When it comes to *evolutionary fears,* as well as to many other areas of learning, animals and people learn by watching what other animals or people do, not by doing something themselves and learning from the consequences. I have the impression this lesson hasn't quite been absorbed by most educators. You read that hands-on learning is best, but that may not always be so. Obviously evolution has selected for strong observational learning in animals and in humans. One of the most amazing examples of this is in Frans de Waal's book *The Ape and the Sushi Master.* Dr. de Waal says that in Japan, apprentice sushi cooks spend three years just *watching* the sushi master prepare sushi. When the apprentice finally prepares his first sushi, he does a good job of it.[21]

Dr. Mineka's research shows how people and animals can develop phobias without ever having had a bad experience with the thing they're afraid of. Classical learning theory always assumed people learn phobias through direct experience. That's logical, but it doesn't correspond to reality, because lots of phobic people can't remember any initial bad experience. Probably most people with fear of flying, just to give a common example, have never come close to crashing.

So a lot of therapists had suspected that phobias are contagious, that people can "catch" a phobia by hanging around people who already have it. Dr. Mineka's research showed that not only is it possible to learn a phobia by being exposed to someone else who has that phobia, it's incredibly natural and easy to acquire a phobia this way. Fear is contagious.

The fact that animals learn what to be afraid of from watching other animals is another example of evolution giving animals and people an ability to ward off trouble before it happens. If you're Mother Nature and you decide to set things up so everybody learns what to be afraid of through direct, hands-on personal experience, you're going to lose a lot of animals. The only monkeys you'd have around to propagate the species would be monkeys who'd had the good luck never to meet up with a snake in the first place, or monkeys who did meet up with a snake and lived to tell the tale. The odds of keeping monkeys on the planet are going to be a lot higher if you set things up so monkeys learn about snakes from other monkeys.

AN ELEPHANT NEVER FORGETS

Of course, it's not going to be much use learning about snakes in the safety of your monkey community if you don't remember what you know the next time you run into one. What happens if your monkey elders tell you snakes are bad news, and it slips your mind?

When you think about how much stuff you've forgotten in your life (quick! what's the quadratic equation?) it's kind of horrifying to think that our survival depends on *remembering* all the bad stuff we're supposed to be afraid of.

Evolution solved that problem by making fear learning *permanent*. All intensely emotional learning is permanent. That's why you can forget everything you ever learned in trigonometry, but no one born before 1958 is ever going to forget where they were when Kennedy was shot, and no one born before 1996 is ever going to forget where they were on September 11. You couldn't forget where you were even if you wanted to, and even if you tried to.

The story is a little different with lesser traumas and fears. Animals and people certainly act as if they can forget a milder fear, and in the past behaviorists did quite a bit of research on this. Typically researchers would teach an animal to be afraid of something neutral, such as a light or a tone; then teach the animals to stop being afraid of the light or tone. They did this by pairing the *conditioned stimulus,* which was the light or the tone, with something aversive, like a shock to the foot or a puff of air to the eye.

Under those conditions, pretty quickly an animal would start reacting fearfully to the light or the tone, at which point the experimenters stopped pairing the light or tone with anything bad. Sure enough, after a while the animals stopped reacting badly to the light or the tone. Behaviorists called this phenomenon *extinction,* because they had extinguished the response. The animals seemed to have forgotten that lights or tones were scary. Researchers found the same thing in humans.

However, it turns out that extinction doesn't actually wipe out the fear from your brain. It's still there. If you teach an animal to fear a tone that precedes an air puff to the eye, and then teach him not to fear the tone because there's no more air puff, he hasn't forgotten. He stops blinking reflexively every time he hears the tone, but all you have to do to get him blinking again is to pair the tone with the air puff again just *once* and the animal is right back where he started. He *knows* that tone means air puff. He hasn't forgotten.

Both animals and people can "get over" a learned fear. But today we understand that getting over a fear isn't the same thing as forgetting a fear. *Extinction* isn't forgetting; it's *new learning that contradicts old learning.* Both lessons—tone is neutral and tone is bad—stay in emotional memory.

FAST FEAR, SLOW FEAR

When you spend a lot of time with animals it's easy to see that animal fears are worse than human fears a lot of the time. It's also easy to see that you, as a human, share certain core fears with animals.

Cows don't like snakes, and neither do you. You and any cow you meet see eye-to-eye on that one.

But beyond that, it's hard for people to empathize with an animal's fears. A lot of times it's hard even to know what an animal's fears *are*. I get a lot of calls from people who can't figure out what's getting their animals so upset. I'll go out to a plant that's having problems and find the manager standing there in the middle of what looks like a perfectly normal, perfectly safe feedlot to *him,* and he's got a couple hundred head of cattle having conniptions. He has no idea why.

To understand animal fears it pays to know something about the brain. One of the most important researchers in the neurology of fear is Joseph LeDoux of New York University. In his book *The Emotional Brain,* Dr. LeDoux explains that fear happens in the amygdala.[22] What's really interesting for nonscientists is that he's found there are two kinds of fear in the brain: *fast fear* and *slow fear,* which he calls the *low road* and the *high road.*

The high road gives you slow fear for a simple reason: its physical path through the brain is longer than the low road. On the high road, a scary stimulus, such as the sight of a snake in your path, comes in through the senses and goes to the thalamus, located deep inside the brain. The thalamus directs it up to the cortex, at the top of the brain, for analysis. That's why Dr. LeDoux calls slow fear the high road. The information has to travel all the way up to the top of the brain. When it gets there the cortex decides that what you're looking at is a snake, then sends this information—it's a snake!— back down to the amygdala, and you feel afraid. The whole process takes twenty-four milliseconds.

The low road takes *half* the time. Using the fast fear system, you see a snake in your path, the sensory data goes to your thalamus, and from there it goes directly over to your amygdala, which is also located deep inside the brain, in the temporal lobes at the side of your head. The whole process takes twelve milliseconds. Dr. LeDoux calls fast fear the low road because the sensory information doesn't have to travel up to the top of the brain. The cortex is out of the loop.

Both systems operate at the same time, with the same sensory inputs. This means that the thalamus receives potentially frightening sensory data and sends it two places: both to the cortex *and* to the amygdala. If you're looking at a snake, the fast fear system has you jumping out of the way in twelve milliseconds; then, twelve milliseconds later, you get a second jolt of fear from the exact same information finally arriving at the amygdala after having been routed through the cortex for closer analysis.

Dr. LeDoux thinks the reason our brains are set up to work this way is that evolution couldn't put both speed and accuracy into the same system. The fast road, he says, is quick and dirty. You're walking down a path, you see something long, thin, and dark in the path, and your amygdala screams, *"It's a snake!"* Twelve milliseconds later your cortex has the second opinion: either, *"It's definitely a snake!"* or, "It's just a stick." That doesn't sound like very much time, but it makes all the difference in the world to whether you get bitten by that snake or not, assuming it is a snake and not a stick. The reason fast fear can be so fast is that accuracy is sacrificed for speed. Fast fear gives you a rough draft of reality.

It's the cortex that does the precision rendering of the world, so it's the cortex that can tell a snake from a stick. But that takes time, and time is exactly what you don't have when you're looking at a snake. Dr. LeDoux thinks nature evolved this system because it's better to be safe than sorry: it's better to mistake a stick for a snake than to keep walking toward a snake bite while your cortex is still forming an opinion.

The other thing to know is that high road fear is conscious, while low road fear is not. High road fear is conscious because it's been through the cortex, which makes you consciously aware of what's scaring you. *I'm scared of that snake sitting there in the middle of the road.* That's conscious, high road fear. With low road fear you react unconsciously, or mindlessly. You're running away before you know what you're running away *from*.

WEIRD FEAR

One of the really interesting things about memory is that *conscious memory* is much more fragile than *unconscious memory*. The terminology for different kinds of memory gets confusing, partly because different fields use totally different terms for conscious and unconscious memory. Some fields talk about *declarative* versus *procedural*; other fields talk about *explicit memory* versus *implicit memory*. I'll mostly stick to conscious and unconscious, but when it makes sense to use other terms, I will.

Conscious memory handles the kinds of things we call "school learning," facts, figures, dates, names, and so on. If you think about how much of what you learned in school you've forgotten you'll get a good idea of how fragile it is. Unconscious learning is *much* more stable and long-lasting. The old saying about how you never forget how to ride a bicycle is a perfect example. It's true: you *don't* ever forget how to ride a bicycle once you've learned.[23] You can have significant brain damage from a stroke, and you're *still* likely to remember how to ride a bicycle. It's very tough to wipe out unconscious memory.

By now you're probably thinking Freud was right. If so, you're not far off. A number of Freud's ideas are turning out to be pretty good descriptions of how the brain works. I'm no expert on Freud, so I should add that I have no idea whether Freud's idea of repression will be supported by brain research. What *is* supported is the idea that we have a huge amount of unconscious information stored up in our brains.

I don't know whether unconscious, or procedural, learning like bicycle riding is always permanent. The easy way to remember what procedural memory is, is to think of things like bicycle riding as *procedures*. When you learn something like riding a bicycle, or how to button and unbutton your shirt, you're using unconscious, procedural memory. Your fingers know how to unbutton your shirt; you can do it without thinking about it consciously.

I don't know whether procedural learning is always permanent, but it looks like fear learning is. Learned fears are the exact opposite of learned facts, dates, and names, which you're constantly forgetting. You never forget a fear. In fact, fear learning in animals and people is so powerful it can get stronger over time, even when you do nothing further to "practice" your fear through repeat exposure. Say you see a snake in the road just once in your life, and it scares you half to death; you could never see a live snake again yet *still* get more and more frightened of snakes as time goes on.

According to Dr. LeDoux, the relative weakness of conscious fear memory compared to unconscious fear memories may explain why fears can spread so far beyond their original content. What may happen is that as time passes you lose your conscious memory of the thing that frightened you, but *your unconscious memory is as strong as ever.*

Dr. LeDoux gives a nice example of a person in a bad car crash where the horn gets stuck on. For a period of time after the crash, the person feels frightened all over again every time he hears a horn. But then, over time, he gradually forgets about the car horn, because the details of the car crash are fading out of his conscious memory. He doesn't consciously remember he's afraid of car horns.

But as far as his unconscious emotional memory is concerned, the crash and the stuck horn could have happened yesterday. Now, whenever he hears a honking horn, his body tenses up and he feels scared, but he doesn't know why. So his conscious mind associates his bodily fear reactions with whatever perfectly innocent things are going on around him, like walking down a busy street, or trying to find the elevators inside a crowded mall parking lot. It could be anything at all. Having forgotten what he was originally scared of, he's developing all kinds of brand-new, totally irrational fears that aren't based in anything real.

In Dr. LeDoux's view, this is one reason why therapists see so many fears without any obvious cause in their patients. What they're seeing are secondary *downstream* fears that developed after the conscious content of the original fear was forgotten. The new fears are like stand-ins, or substitutes, for the old one. This may sound strange, but it happens a lot, especially to people with phobias. As Dr. LeDoux says, "phobics can sometimes lose track of what they are afraid of." [24]

Another thing that could happen, once the conscious details of the original frightening experience have faded, is that a person can start having conscious feelings of fear that aren't attached to anything he can pinpoint. They just seem to come out of nowhere. Say he hears a honking horn somewhere in the distance. He doesn't pay any attention to it and then starts to feel anxious without realizing it's the horn that caused the emotion. His conscious memory has forgotten all about the horn, but his amygdala hasn't, and he could end up thinking of himself as an anxious person.

Dr. LeDoux thinks the differences between the fast fear and slow fear systems probably lead to lots of the different anxiety disorders psychiatrists treat. As he points out, the slow fear system is probably the reason a person develops a fear of a harmless car horn in the first place. The stuck horn didn't cause the car crash; the car crash caused the stuck horn. But the amygdala doesn't make the distinction, and everything about the scene of the accident can become contaminated with fear. All kinds of irrational fears probably develop because the amygdala reacts so fast based on such crude analyses of a situation.

This process happens to animals all the time. I got a call to work with a horse who was terrified of garage doors. When I talked to the owners I found out that the first time they tried to collect semen from the horse, he'd fallen on his butt. To collect semen you have the horse mount a dummy, and somehow this horse had fallen backward. It was a freak accident, and the people working with him got crazy and hit him with the whip and yelled at him, so now he was traumatized.

The reason he was terrified of garage doors was that he'd been looking at a garage door when he fell. The garage door had nothing to do with the fall, but his amygdala made the crude association: garage door–traumatic fall.

The next time they tried to breed the horse they kept him out in the open away from any buildings and he was fine. But a horse who's going to go berserk anytime he sees a garage door is a dangerous horse to ride or handle anywhere outside his home corral.

ANIMAL FEARS ARE DIFFERENT

Although the basic fear mechanisms in an animal's brain are the same as in a person's brain, the difference in frontal lobe size and complexity means that animal fears and human fears end up being different.

The single most important thing to remember is that animals are afraid of tiny details in their environments. I like to use the term *hyper-specific* to describe animal fears. It comes from autism research, because *autistic people are extremely hyper-specific*. It's one of the main things that separate them from typical people. I'll be talking more about hyper-specificity in autistic people and in animals when we get to animal genius, so for now all I need to say is that being hyper-specific means you see the differences between things a lot better than you see the similarities. You see the trees better than the forest. A lot of times you might not see the forest at all. Just trees, trees, and more trees. Animals are like that.

My favorite example of a hyper-specific fear is the black hat horse. I met the black hat horse when his owner came to me for a consultation. She said her horse was terrified of people wearing black hats. That was all, just black hats.

Now that is an extremely specific fear. It was so specific I was amazed a normal human being had managed to figure it out. It might seem easy to notice that a horse is bolting every time he sees a black hat, but it's not. If you think about it logically, there's almost an infinite amount of data coming into our senses every second of the day. The only reason the world isn't a total blur is that your nervous system automatically filters out a huge amount of stuff, and automatically focuses on some things and not others. That's what inattentional blindness is all about, filtering out the stuff you don't care about.

Normally, a typical human nervous system is not set to focus on black hats or any other extraneous detail. It's just not. But an *animal's* nervous system is set to focus on detail, because his frontal lobes are so much less developed than a typical human's frontal lobes. That's why an animal can become terrified of black hats: (a) because he notices them in the first place, and (b) because he has less frontal lobe power available to analyze and suppress a fear of black hats once the fear gets going.

I was impressed that the horse's owner had managed to figure out that black hats were the problem. She had managed to see through her horse's eyes, and the ability to do that is rare.

She and I worked with the horse together. We wanted to know two things: what were the exact parameters of his fear, and could we train him out of it? We found out pretty quick that he was really focused on that black hat. We tested him on all the hats we had between us: a red baseball cap, a light blue baseball cap, and a white cowboy hat. The only thing that bothered him was a black cowboy hat, and it had to be black.

He was so scared of the black hat that I didn't even have to be wearing it to set him off. If I stood perfectly still in front of him just holding a black hat quietly at my waist, he would start to rear. He was looking straight at me, but the only thing he was taking in was the hat. That made me bad. He was sensitive to the position of the hat, too. The closer I held it to my head, the more trouble he had.

So the problem was the black hat, and only the black hat. After that we tried to desensitize the horse. When it comes to fear, there are only two techniques that work with animals at all, and neither works very well: *desensitization* and *counter-phobic training*. Desensitization is exactly what it sounds like. You expose a person or an animal to tiny doses of the thing he fears, building up gradually to larger and larger doses. Counter-phobic training means pairing the thing an animal or person fears with something he likes, such as food. You're trying to build in some good associations to counter the bad associations.

We did a long session of desensitization with the black hat horse, and we made some progress. By the end I was able to have the owner put the black hat on the ground, and I could lead the horse up to it. We even got him to touch it with his nose. But that was as far as we could go.

That is a classic example of the kind of hyper-specific fear animals develop all the time. The horse's category for *bad* and *scary* was *black hats on people*. Not white hats, not red hats, not blue hats. Just people wearing or holding a black hat, although he wasn't exactly keen on the sight of a black hat lying on the ground, either.

You see this all the time with animals. I met a ferret once who was afraid of the sound of a nylon ski jacket. Someone wearing one had probably abused him, and what he focused on was the sound of the person's jacket. So that's what set him off, the sound of nylon swishing against nylon. Another time I went to a zoo where the keepers told me their chimpanzees were terrified of burlap cloth. They'd been tied up inside burlap bags after they were captured, and if you put a piece of burlap in their cage they'd immediately bury it under the straw, out of sight so they couldn't see it. Then they all felt a lot better.

BEING HYPER-SPECIFIC

It's extremely important to understand how hyper-specific animals are, because you won't socialize your animals properly if you don't. I've watched animals at meatpacking plants go berserk when they saw a man on foot for the first time in their lives. Up to that point the only men they'd ever seen were men riding horseback. These were beautifully handled animals who'd been worked with quietly and gently, but when they saw a man on foot they panicked and almost trampled him. The mental category they'd formed was man-on-horseback, or maybe just man-horse, like a centaur. They didn't automatically expand their man-on-horseback-is-safe category to include man-on-foot-is-safe.

Another example. Richard Shrake, the famous horse trainer who developed resistance-free training, says it's important to train a horse to let you mount him either from the left side *or* the right. You have to do that because to the horse these are two completely different things. A horse that suddenly has to be mounted from the right when he's always been mounted from the left could buck or bolt. It's dangerous.

Same thing with dogs. I had an interesting talk recently with a lady who keeps wolf hybrids for pets, something I don't recommend. She told me that if you're going to have a wolf hybrid as a pet, you *have* to socialize it between four to thirteen weeks of age that *all* men are okay, not just the male owner. Otherwise they'll think the owner is okay, all other men are the enemy. You have to do the same thing with women, children, toddlers, and babies, and you have to socialize the animal to different members of each category *separately*. It's not just the owner's little toddler who's okay, *all* little toddlers are okay. It's not just the owner's wife who's okay, *all* women are okay. And so on.

Another way to think of this is that animals don't generalize well. They don't generalize from male-owner-is-okay to male-owner-and-mailman-are-okay. Normal human beings are almost exactly the opposite: normal human beings tend to err on the side of over-generalizing, not under-generalizing. That's what a stereotype is, an over-generalization. All women are X or all men are Y. It's natural for normal humans to think that way, but you have to actively *teach* an animal to group all women inside the category "women." (Animals do form categories, which is a kind of generalization. We'll get to that in the next chapter.)

I find that even people who work with animals professionally don't tend to pick up on this aspect of animal minds. It's just too foreign to their own way of processing the universe. Even when a trainer or handler gets pretty good at analyzing what's scaring an animal it's still hard for a normal person to get a sense of animal emotions. What's it like being so vulnerable to tiny details?

Even though I'm fairly hyper-specific myself, I don't know the answer. But I *think* it has something to do with fear of the unknown.

Fear of the unknown is universal. Everyone has *some* fear of the unknown, although of course people also like novelty and variety within limits. Animals do, too. They're afraid of the unknown, but they're also drawn to it.

If you think about it, animals are constantly confronting the unknown. For an animal who's never seen a man off a horse, *a man walking on his own two legs is an alien.* So I think a good way to try to get inside an animal's head, to the extent that's even possible, is to be constantly asking yourself, "How would I feel if what I were looking at right now was something I'd never laid eyes on before in my life?"

A friend of mine came up with an analogy to the cattle who panicked when they saw a man walking on two feet. "If I were sitting in my living room reading a book," she told me, "and I looked up and saw a stranger walking down the sidewalk and up to my front door *on his hands,* acting as if there was nothing out of the normal going on, I'd probably be scared half to death." She said it gave her the creeps just thinking about it.

That would probably scare anybody. When you see something you've *never* seen before, something you never expected to see, you're going to feel some fear. That's because we're wired for survival, so when we confront the unknown our survival brain gets activated and starts screaming at us, "What is it!? What is it!?" And, *"Is it dangerous?!"*

FEAR AND CURIOSITY

I talked about cows being curiously afraid in Chapter 3.

What's interesting about animals being curiously afraid is that it's the most fearful animals who are also the most curious. You'd think it would be the exact opposite. A fearful prey animal like a deer or a cow ought to just get the hell out of there whenever it sees something strange and different that it doesn't understand.

But that's not what happens. The more fearful the animal, the more likely he is to investigate. Indians used this principle to hunt antelope. They'd lie down on the ground holding a flag, and when the antelope came up to investigate they'd kill it. I've never heard of Indians lying down on the ground holding a flag to catch buffalo, and my bet is that's because they never did it. Buffalo are big-boned animals, and we know for a fact big-boned animals are less fearful than animals with small bones. I'm guessing, but I don't think a buffalo is going to be as compelled to investigate a flag flying in the middle of the prairie as an antelope is, because he's not as fearful as an antelope is. He's a great big strong buffalo; what does he have to worry about? But a delicate little antelope has a *lot* to worry about, and that's why he's always looking into things.

You see the same difference in horses, too. Arab horses are fine-boned and flighty, while Clydesdales are calm. If you put Arab horses together with a bunch of Clydesdales, and hang a flag on the fence, it's the Arab horses who'll walk up to the flag first. The Clydesdales will always be the last. Curiosity and fear go together.

Fear *seems* to correlate with intelligence, too, although no one can say that for sure. I mention this because any horse trainer will tell you Arab horses are the smartest. If we were to find out that high-strung animals are more intelligent than placid animals, the difference may be due to the fact that nervous animals investigate their environments more, learn more, and get smarter in the process.

THE NEW NEW THING

I think what all of this means is that animals probably spend a lot more time being suddenly exposed to something brand-new they've never seen before than humans do. First of all, animals have more limited lives than people do, if only because they don't read books and watch TV. They haven't had the huge amount of vicarious experience we have. Most of us have never seen a pyramid in Egypt, but we wouldn't be shocked if we did, because we've seen the pyramids in pictures.

But second, animals' hyper-specificity also means they're constantly coming face-to-face with new things they haven't seen, heard, touched, smelled, or tasted before. If you're hyper-specific and you've seen a few big dogs in your life, but you've never seen a dachshund, then a dachshund doesn't automatically seem like a dog the first time you do see one. We don't know *how* hyper-specific animals are, but we do know they're a lot more hyper-specific than nonautistic humans are. I think that probably has to mean that animals encounter more new things than people do, if only because people automatically assign most new things to old categories.

That's why seeing a dachshund for the first time when I was little completely threw me off—because I'm hyper-specific. To me, that dachshund was brand-new, whereas to a nonautistic person it would have been just another dog.

HOW AN ANIMAL'S FEARS GROW

Animal fears spread like crazy.

People's fears spread, too, as I mentioned, but *animal fears spread in a hyper-specific way*.

Here's my best example. Mark's dog, Red Dog, is deathly afraid of hot air balloons. She starts going crazy when a hot air balloon is just a tiny speck miles away in the sky.

We have a lot of hot air balloons in Colorado, and originally Red Dog got spooked when one of them revved its burner right over her house. Since that one bad experience she's gotten more and more frightened of the balloons, exactly the way Dr. LeDoux describes. Her fear has gotten stronger, not weaker, and it's spread out to all other hot air balloons, no matter how far away.

People's fears can grow that way, too. But now Red Dog is branching out in a way I don't think people do. Just lately she's gotten terrified by the sight of those red aerial marker balls they put on power lines so airplanes won't hit them. She goes nuts when she sees one of those things.

Then the other day all of a sudden she went crazy when she saw the rear end of a gasoline tanker.

I hadn't given much thought to Red Dog's choice of objects to be terrified of until I reread Dr. LeDoux's book. Halfway through I suddenly realized that the things Red Dog is afraid of are just different versions of the same thing: all three of them are round, red objects seen against the blue sky. The tankers are round and painted red on the back, and Red Dog sees them when she's riding with Mark in his truck. From her angle, she's probably seeing them surrounded by sky.

When human fears spread from the original scary thing to other objects or situations that should be neutral Dr. LeDoux calls it *overgeneralizing*. The fear generalizes too far. A Vietnam vet who jumps out of his skin when he hears a car backfire is over-generalizing from the sound of gunfire to the sound of cars backfiring.

That's what Red Dog was doing, but she was over-generalizing in a hyper-specific way.

People can make hyper-specific over-generalizations, too. That's what the Vietnam vet is doing when he jumps at the sound of a car backfiring. But animals do it all the time. I don't think any human would go from being scared of red hot air balloons to being scared of the red ends of tanker trucks.

Animals seem to over-generalize within the sensory channel that first frightened them. That's why Red Dog keeps generalizing out to things she can *see*. People probably do this, too, but my impression is that people's over-generalized fears are often more logical and more *conceptual* than an animal's. For instance, I've heard of people going from fear of flying to fear of elevators. That's different from a hot air balloon fear spreading to aerial markers. If an elevator crashed with you inside, that would kill you just as surely as a plane crash would, but no aerial marker is going to rev up its burner over your house and startle you half to death. An airplane and an elevator are linked *conceptually*; a red hot air balloon and a red aerial marker are linked only *perceptually*.

Some of the difference between animal fears and human fears is probably due to the fact that animals know less about the world than we do, seeing as how we built it and they didn't. Red Dog doesn't know the purpose of hot air balloons, aerial marker balls, or liquid nitrogen tankers.

But even if that's true, you always need to keep in mind that animals are going to generalize their fears out to things that are in the same *sensory* category, not the same *conceptual* category. The black hat horse generalized to other black cowboy hats, not to hats in general. (I wish to heck I'd thought to test him with a big black purse, too. I'd love to know whether anything with the general shape of a black cowboy hat would have frightened him.) Animal fears are hyper-specific, and they spread hyper-specifically, too.

KEEPING FEAR OUT OF ANIMALS' LIVES

With animals, just like with people, there's a difference between traumatic fears and plain old everyday fears. Traumatic fears in animals are always bad news; they last forever, and they can spread. Even if you do manage to put together a fairly effective counter-phobic behavior program, you're going to be doing that program for the rest of the animal's life. It's a lot of hard work, without a lot of gain.

Everyday fears are different. Unless an animal is anxious by nature, an everyday run-of-the-mill fear won't wreck his life or yours, either. The problem is that it's hard to predict which experiences will traumatize an animal and which experiences will just give him something to think about.

Dog owners face this mystery when it comes to deciding whether to install an invisible fence. An invisible fence, for anyone who doesn't know, is a perimeter created by a radio signal broadcast to a receiver the dog wears on his collar. When the dog gets close enough to the perimeter he hears a warning beep; if he ignores the beep and keeps going he gets a shock.[25] You can think of it as a *beep-and-shock* fence instead of a wire fence. Most of the time invisible fences work great.[26] I'd recommend that every dog owner buy one, if I weren't worried about people holding me responsible when they spend anywhere from a couple hundred to fifteen hundred dollars putting in an invisible fence that turns out to be more trouble than it's worth for their particular pet.

The reason some dogs don't do well with an invisible fence relates to pain levels as well as fear levels. A low-fear, low-pain dog like a retriever, either golden or Labrador, can sometimes just run through them. I knew one family whose golden retriever would bound through the perimeter on his way out of the yard but then refuse to come through it on the way back. He didn't want to get shocked. Apparently he didn't mind getting shocked when he was making his Great Escape, but getting a shock just to come home again wasn't worth it.

It was a huge nuisance, because there was one family down the street who was terrified of that dog, even though he'd never done anything bad to them. Naturally that was the one house he'd always make a beeline for whenever he was done with his travels. He'd plop himself down on their doorstep and just lie there waiting for his owners to come get him and take him home. Probably he'd noticed that his owners always seemed to show up the fastest when he landed at the scared family's house. That was true, of course, because the instant the scared family saw the dog they'd start frantically calling the owners every five seconds—and naturally the owners would race over to retrieve the dog the minute they got the call, because they knew how upset the scared family was. Until the owners arrived, the scared family would be locked up inside their house, too terrified to come out. Naturally the owners lived in fear of having this happen sometime when they weren't home. What if there was an emergency and the scared family was trapped inside their house because the dog had busted through the invisible fence again?

I heard about another dog, a little Jack Russell terrier, who would get through the fence just because his fellow-dog, another retriever, could go through it. The retriever would sail through unscathed, and the Jack Russell would lower himself to the ground and stare at the place where he knew he'd get the shock. Finally he'd bolt. The lady who told me about him said, "He'd decide to take the hit." I'm sure if that dog had lived alone, or at least in a house whose other dog wasn't a retriever, he would have stayed put. But he wasn't going to let his pal take off without him.

Those are the problems you can have with dogs who are low-fear (or low-pain). They're unusual, but they do happen. The problems that can crop up with a high-fear dog are more difficult to manage. I've never heard of a dog getting out-and-out traumatized by an invisible fence, but I've seen some come close. Some dogs will get so scared of the perimeter that they'll refuse to ever go through it, whether the collar is on or off, and including when you put them on a leash to take them for a walk. You have to carry or drag them through the perimeter.

That's not so horrible, but I also heard about a two-year-old collie who got so scared of her own yard that she lost her house-training and started pooping inside the house. If her owners would force her to go outside she'd just stand on the deck barking until her owners finally gave up and let her back in. Then she'd poop on the carpet.

These are all unusual cases. Most dogs live happily inside an invisible fence and don't panic when you walk them through the perimeter on a leash. But even when an invisible fence works perfectly, you still have to keep on top of the situation. Although animal fears, like human fears, are permanent, animals *will* reality-test a fear that falls short of a phobia.

I know that happens with invisible fences. I talked to a woman who bought an aboveground invisible fence for her two young dogs. It worked like a charm, but remembering to put their collars on every morning was a pain. (She didn't like the dogs to sleep in the collars at night, because one of them had sensitive skin and the metal prongs were rubbing it raw.) So she figured she'd be vigilant for a couple of months until the dogs took it for granted that they couldn't leave the yard without getting a shock. Then she wouldn't have to worry about whether one of the dogs got out of the house without the collar on. She said she based this on some story she read back in college about how B. F. Skinner once trained some sheep to stay inside a fence, then replaced the fence with a symbolic wire strung between posts. Supposedly the sheep never tried to get past the wire, even though they easily could have.

I don't remember ever seeing that story in Dr. Skinner's work myself, and I'd be surprised if that's what he found. In my experience some animals don't test fences, but others do. That lady turned out to have fence-testing dogs. At first everything seemed to be working out. The dogs never went near the boundaries, whether they were wearing their collars or not. They didn't act like they associated the shocks with the collar, either, because every time she took their collars off to take them for a walk she'd have to pull them through the perimeter. They were scared of getting a shock whether they had the collars on or off.

So after a while she just stopped worrying about getting the collars on first thing in the morning. Big mistake. One morning she was sitting outside reading the newspaper when she noticed the dogs running a couple of feet up the hill beside her house, then coming back down again. They seemed to be doing this repeatedly, although she wasn't paying close enough attention to be sure. She thought they were getting awfully close to the shock perimeter, but since she figured they'd been permanently conditioned like Dr. Skinner's sheep, she didn't worry about it.

The next thing she knew, both dogs were gone. They stayed away for hours and probably had a nice romp around the pond a little ways from her house. She's been having problems with them ever since. As long as she has the collars on and the batteries are working, they stay home. But if she slips up—either forgets to check the batteries or slacks off on putting the collars on in the morning—it doesn't take too long for the dogs to figure out they're free.

I don't know how they manage it, but it sounds like they're doing their own doggie version of reality testing. The owner has observed that every time she forgets the collars for a few days the same sequence unfolds. First the dogs stay well within the invisible fence boundaries, collar or no collar. Then they start expanding the perimeter, going a little bit farther than the collar would let them go, but no farther. Then, not too long after that, they're gone.

What she couldn't figure out was, how do the dogs know it's okay to expand the perimeter? They're still acting scared when she takes them through the perimeter on a walk, so why do they test it on their own?

I think they are probably picking up signals a human can't perceive. I'm guessing they get some kind of little vibration or early warning buzz from the receiver *before* they reach the spot where the warning sound beeps. They get a warning before the warning. Once the dogs stop perceiving the pre-warning sound or sensation, they start testing the boundaries.

The reason I think this is that the dogs *never* set off the warning beeps. That has to mean that somehow they know it's safe to start pushing out the boundaries. If they were just sporadically testing from time to time, to see whether the perimeter was still there, they would set off beeps on days when their collars are on, which is most days.

However those two dogs are doing what they're doing, the Mark Twain saying about the cat on a hot stove is true only as far as it goes. "She will never sit down on a hot lid again—and that is well;" he said, "but also she will never sit down on a cold one anymore." That's true only of a cat who got burned badly enough to be traumatized by the experience, or of a cat who didn't get burned too badly but doesn't have any good reason to sit on the stove apart from the fact that cats like to be up high. If the cat isn't flat-out terrified of the stove, just leery, and if there's a plate full of yummy meat sitting up there, I predict most cats are going to be back up on that stove.

Fear Monsters

Temperament is everything. An animal with too much fear by nature—or too little fear—can be hard to live with and manage. Owners and trainers have to match their approach to temperament. The wrong kind of handling with large prey animals like cows and horses can actually make them dangerous. You can take a perfectly normal horse or cow and turn it into a spin-and-kick animal—an animal who will spin around and kick a human being with both hooves. When that happens you've taken a prey animal and turned him into a killer. It's ridiculous.

You see it happen when owners use rough training to teach a horse or cow to accept a halter and lead rope. They put a real strong halter and a six-foot lead on the animal, tie him up to a pole, and let him fight it out with the post until he's exhausted and gives up. The owner is trying to teach the animal to walk calmly on a lead, but instead of just putting on the halter and lead and letting the animal wear them around the corral to get used to the feeling, they think they have to break the animal's resistance.

It's a horrible training method. But it has different effects depending on an animal's temperament, especially his level of inborn timidity. Calm animals, like Holstein cattle, will habituate. After rearing and bucking for a while they'll settle down and get used to the situation. It's still a stupid way to train them, but they can take it. A more sensitive, fearful animal can become scared, skittish, and unmanageable when you try to train him that way. That animal will *never* be okay with the halter and lead, for the rest of its life.

But it's the animals with the medium temperaments, in between calm and fearful, who become dangerous. When you tie them up to the post they get scared and stay scared, *but they don't lose control.* They're the ones who learn to spin and kick. A naturally calm animal like the Holstein doesn't care enough about being tied up to a post to *need* to learn to spin and kick, because he doesn't feel that his survival is threatened. Naturally timid cattle do feel that their survival is at stake, but they get too panicked to do anything about it. It's the in-between animals who have exactly the right amount of terror they need to learn how to kill a human being. After rough training to the halter and lead they've learned that they have two cannons for back hooves.

—

I call high-fear cattle fear monsters, because they get completely overwhelmed by panic. I've seen Saler cattle (Salers are French dairy cattle we use as beef cattle) get so frantic they'll fall on the ground and start rolling around. A Saler cow who gets her leg caught between the loading dock and the truck can actually rip her own leg off just below the knee in panic. I saw this happen one time. It was horrible. An Arab horse can do the same thing. These animals are fear monsters. They get so terrified they destroy themselves.

A couple of good things about Saler cows, though: they're excellent foragers, and they're wonderful mothers. Saler cattle are dairy cows who were developed in the French mountains, and they'll go anywhere to find grass. They'll climb up into nooks and crannies a fat old Hereford wouldn't think of even trying to get to. And they'll fight off anything that threatens their calf; they'll fight off a coyote every time. Of course that means they'll fight you off, too, if you try to do anything to the baby. So you have to be careful.

Holstein cows, on the other hand, are so calm now they're terrible mothers. They've been selectively bred to be calm and to be huge milk producers, and we've bred their protective maternal instincts right out of them. If a coyote really wants their calf, the coyote can have him. Nothing gets a Holstein excited. Meanwhile Holstein bulls can be dangerous because they have no fear.

IS IT BAD BEHAVIOR OR IS IT FEAR?

A big problem I see with a lot of trainers and owners is that they don't know when an animal's bad behavior is motivated by fear. I knew a dog with fear-based aggression who, when her owner took her out for a walk, would start barking like crazy anytime anyone came near. The dog was barking because she was scared, but the owner didn't understand. When the dog kept barking and ignoring her owner's commands to "hush," the owner would start getting upset herself and would eventually start yelling at the dog. That made things worse, because the dog thought her owner was screaming for protection, so she got even more crazed.

In that case the owner was lucky, because she figured out what was going on before too much damage had been done. Once she realized the dog was being aggressive because she was scared, she started a whole new program. One of the things she did was that anytime a bicycle rode by she'd stop walking and have the dog sit down. She'd stroke her and talk quietly, telling the dog everything was okay. She was able to get a lot calmer behavior out of her dog that way. (Bicycles were especially hard because not only is a bicycle something that's being ridden by a *scary stranger*, it's in motion, and that sets off a dog's natural drive to chase moving objects.)

I mentioned earlier that I'm not a big fan of punishment as a teaching method no matter what an animal's temperament, except in the case of prey-drive-motivated chasing of joggers and bicyclists and the like. But punishment is worse for some animals than for others. There are calm animals who can deal with punishment just fine, and there are nervous animals who totally fall apart if they experience a lot of anger from their human owners.

You have to match your handling to the animal. High-fear animals need super-gentle handling. Low-fear animals don't need harsh handling, but they don't fall apart if they get it. I saw some Paso Fino horses down in Argentina who could take just about anything their owners dished out. The trainers really abused them. They beat the horses into submission, and they put wires attached to *tie-downs* around their noses. A tie-down is a short strap on either side of the horse's face that is attached to the girth of the saddle. Normally the tie-down is *loosely* fastened to a broad leather strap that goes across the horse's nose. People use tie-downs to keep a horse from tossing his head, and some trainers think tie-downs keep a horse from rearing. But tie-downs make horses crazy, so there's no reason to put one on tightly and there's certainly no reason to attach it to a wire that would cut the horse's nose.

Every one of those horses had a quarter-inch dent in its nose. If you did that to an Arab horse, he'd be crazy and unrideable for life. The Paso Fino horses are low-fear, and they habituated—but they hated people. The minute I touched their forelocks, they pinned their ears back and bared their teeth. That was as far as it went, because they knew they'd be beaten if they bit me. But there's no good reason to make a horse hate humans that way.

Some trainers swear rough handling is effective. But what's interesting about these trainers is that if you check out their horses, they're all big-boned, low-fear horses who habituate fast to treatment that would crush a high-strung animal. Mark noticed this one time at the racetrack. The rough trainers were all working with big, heavy horses, and they all think Arab horses are crazy. The gentle trainers were working with the fine-boned, nervous animals.

BRINGING UP BABY

A while ago I read an article about the Homeland Security alerts that had a good line in it: "Once you scare people, it's hard to un-

scare them." Since it's just about impossible to un-scare a seriously
scared animal, you should do whatever you can to fright-proof your
animals.

That means, first of all, you have to expose any pet or animal you
own to other animals and other people he's likely to come across—
and you need to do this *when the animal is young*. I've already talked
about how important it is to socialize animals to other animals and
other people, in order to prevent them from developing aggression.
But it's also important to expose them to other animals and other
people to prevent them from developing hard-to-manage fears.

If you own a riding horse, you should train him to be as comfortable with novelty and change as possible. You can introduce novelty
into a grazing animal's life by doing things like tying a yellow raincoat to the fence one day, or having him close by when you raise the
hood of your car. It can be anything. You're trying to get him to
expect the unexpected, or at least not go ballistic when the unexpected happens.

It's easier to do this when an animal is young and you can just have
it trail along after its mom. If the mother isn't afraid of the new things
you're showing the calf, the calf won't be, either. (This is what Dr.
Mineka found in her research with lab-reared monkeys and snakes.)

The fact that animals can be inoculated against fears by other animals is something your vet probably won't think to mention. There
are two sides to this coin. First, when you get a new pet you have to
be careful about the other animals he meets in the beginning. I
know a situation right now, a couple with two Pomeranians they got
at different times, that's shaping up to be really depressing because
the first dog is teaching the second dog all the wrong lessons.

The first Pomeranian, who was around two years old when they
got him, was scared to death of the husband from the minute the
wife brought the dog home. That's not uncommon; a lot of animals
are scared of men, I find. But this dog was so neurotic about the
husband that they think he may have been abused by the teenage
son in its previous home. They've worked and worked with that
dog, trying to get him to relax around the husband, but two years
later he's still scared. When he has to be alone in the house with the
husband he hides out in his crate.

Then a couple of months ago their older dog died suddenly, and they got a second Pomeranian to take her place. This time they made sure the dog didn't have any emotional problems with men or anyone else before they brought him home.

For the first week or so everything was fine. The new dog wasn't afraid of the husband, and he adjusted great. Then almost overnight his attitude changed. All of a sudden the new dog is afraid of the husband, too. The husband hasn't done anything bad to him, but the new dog is scared. So now when the wife's away *both* dogs are cowering inside their crates. It's pretty demoralizing being alone in your house with two dogs who won't talk to you.

I'm sure the new dog learned his fear from the first dog. The only owner he'd had to this point was a woman, so he probably hadn't seen many men, and he hadn't learned that men were okay. Since animals learn whom to be afraid of from other animals, the scared Pomeranian apparently taught the new dog that the husband was someone to fear.

What they should have done was have the new dog and the husband spend some time alone together without the scared dog around to mess things up, preferably in the company of another dog who wasn't scared of men. They needed to inoculate the new dog against husband fear before he got home and learned it from their other dog.

Using animal role models to calm animal fears is an old trade secret in horse racing. In her book on Seabiscuit, Laura Hillenbrand says Seabiscuit was a "train wreck" when Charles Howard bought him; the horse was burned-out and mean. His first trainer said Seabiscuit could run but wouldn't, and he chalked it up to laziness. Seabiscuit's other problem was that he refused to exercise hard enough to get in shape. More laziness. The trainer had dealt with it by whipping Seabiscuit like crazy all through every race, and entering him in more races than horses normally run. He figured Seabiscuit spent so much time resting that he was up to it, and besides, the horse was so intelligent he'd "back off if he became overworked."[27]

It didn't work. Seabiscuit was a medium-type horse by temperament, so being whipped all the time and raced too hard got him just upset enough to make him mean as spit.

His new trainer, Tom Smith, decided right away to pair him up with an animal friend to help "defuse" him. Laura Hillenbrand writes that all kinds of stray animals have lived with racehorses, from German shepherds to chickens to monkeys. Tom Smith picked a nanny goat for Seabiscuit and put her in the horse's stall. You can get a good idea of what a mistake it is to mistreat a medium-fear horse from reading what happened next: "Shortly after dinnertime, the grooms found Seabiscuit walking in circles, clutching the distraught goat in his teeth and shaking her back and forth. He heaved her over his half door and plopped her down in the barn aisle. The grooms ran to her rescue."

The goat was out, so Tom Smith brought in a lead horse called Pumpkin. Pumpkin was a classic low-fear animal; Ms. Hillenbrand says he was the kind of animal horse people called *bombproof.* Pumpkin had been a cow pony in Montana, and "out on the range [he'd] experienced everything, including a bull goring that had left a gouge in his rump. He was a veteran, meeting every calamity with a cheerful steadiness. . . . Pumpkin was amiable to every horse he met and became a surrogate parent to the flighty ones. He worked a sedative effect on the whole barn." Tom Smith used Pumpkin as his general "stable calmer-downer," and that's what Pumpkin was for Seabiscuit, too. The two horses stayed together for the rest of their lives. Pretty soon Seabiscuit also had a dog named Pocatell and a spider monkey called JoJo living with him in the barn, too.

That was the beginning of Seabiscuit's rehabilitation, and it's a principle anyone can use with a flighty animal. You don't need any special training, you just need to find the right match—and remember never to put a nanny goat in with a crazed thoroughbred.

FIGHTING FIRE WITH FIRE

If an animal you own or manage does develop fears that interfere with his life or yours, your next step is almost certainly going to be setting up a desensitization or counter-phobia program. I won't go into those, because there are good books on how to do them, and because books may not be enough. You may need to hire a trainer.

There is one other approach you may be able to try if the circumstances are right. That is to fight fire with fire by using an animal's hyper-specific nature to fight a hyper-specific fear. This is a neat trick I learned from a rancher who bought an abused horse nobody could ride. The horse had been abused with a snaffle bit. Snaffle bits have a joint in the middle that sits on the animal's tongue, so the new owner just put on a different bridle with a single-piece bit, and the horse was fine! (A single-piece bit doesn't have a joint; it's all in one piece.) Here was an abused animal, whose fear memories were permanent, and the owner turned him into a perfect riding horse in thirty seconds just by changing the bit. The horse's fear category was hyper-specific: "snaffle bits are bad," not "all bits are bad." He didn't make the connection between snaffle bits and the single-piece bit. They were two different things

I wish I'd known that years ago. When I was in college my aunt bought me a horse named Sizzler who was fine if you walked him or trotted, but would buck every third or fourth time you pushed him to a canter. She'd picked him up cheap from a dealer, and that was why. Sizzler was too dangerous for me to ride, and my aunt couldn't use him on her dude ranch. You can't have a horse who throws the guests. So we had to sell Sizzler back to the dealer.

If I'd known then what I know now, I would have gotten my English riding saddle from high school, and a different pad, and put those on his back. Sizzler was a Western-trained horse, and he'd always been ridden with a Western saddle. I bet if I'd brought out that English saddle, Sizzler would have been fine. He would have thought the English saddle was a completely different object on his back, and he would have been starting fresh.

The moral of the story is: if an animal in your life has a fear you can solve by completely removing the thing he's afraid of, you're in luck.

CHOOSING A STURDY ANIMAL

Fearful animals tend to be high-maintenance, so if you want an animal who's easy to fit into your life you should select for a calm, non-skittish temperament. That's not too hard to do, although if you're picking out a baby animal there are no guarantees, the same way there are no guarantees with human babies.

I've already said mutts are your best bet. Purebred dogs are being ruined by breeders, including even the good breeders, because when you over-select for any particular trait you always get problems. And, as you can see in the case of the rapist roosters, over-selection for single traits will eventually lead to neurological problems.

There are still some good breeds, and there are always individual dogs belonging to chancier breeds like Rottweilers and pit bulls who are good sweet dogs. But don't let people tell you that Rottweiler or pit bull aggression is a "myth." It's not. *Temperament and appearance are connected.* We don't know much about how they're connected, unfortunately, but we know they are.

My favorite example of the connection between appearance and temperament is Dmitry Belyaev's silver fox breeding experiment in Russia. Dr. Belyaev was a geneticist who believed natural selection determined the traits we see in domesticated animals. Dogs got to be the way they are because dog behaviors helped them survive and reproduce.[28]

To test his hypothesis, he set up a natural selection study using silver foxes. He wanted to see if over several generations he could turn wild foxes into a domestic animal like a dog. So in each generation, he allowed only the most "tamable" animals—the foxes most willing to tolerate contact with humans—to mate.

He started this project in 1959 and when he died in 1985 another group of scientists picked up where he left off. Altogether the foxes have gone through forty years and more than thirty generations of selective breeding for tameness. Today the foxes are very tame, though not as tame as dogs. The researchers say that when these foxes are puppies they compete with each other for human attention, whine, and wag their tails. They're turning into domestic animals, just as Dr. Belyaev thought they would.

What's interesting is that their appearances have changed right along with their personalities. One of the first things to change was fur color: they changed from silver to black and white, liked a Border collie. They look quite a lot like Border collies in photos. Their tails also started to curl up, and some of the foxes developed floppy ears. The floppy ears are neat, because Darwin said there wasn't a single domesticated animal who didn't have floppy ears in at least one country where it was found. I don't think that's true any longer, because I can't think of any breed of horse in any country that has floppy ears, although every other kind of domesticated animal does have at least one or two breeds with floppy ears. The only wild animal I know of with floppy ears is the elephant.

Looking at photographs of these animals, I think their bones also thickened, which is what I would expect given that fine-boned animals are more high-strung. Belyaev was breeding his foxes to be calm, so he probably started getting slightly bigger animals, with thicker bones.

The tame foxes developed brain differences right along with their physical and behavioral differences. Their heads are smaller, they have lower levels of stress hormones in the blood, and they have higher levels of serotonin, which inhibits aggression, in the brain. Another interesting change: the skulls of the male foxes have been *feminized*. Their heads are shaped more like a female fox's head than like a wild male fox's head.

Eventually some of his foxes developed neurological problems, just like you'd expect. They had epilepsy, and some of them started holding their heads back in a strange position. Some of the moms even ate their own puppies. Pure over-selection programs always bring trouble.

I worry about this happening with golden retrievers and Labradors who are bred to have calm temperaments. Recently they've started having some very unusual aggression problems in goldens, and I've had at least one owner tell me that goldens have become hyper animals. She's owned goldens for years, and she always has three or four goldens at the same time, so she's noticed a difference. That's just one person's experience, but what she's reporting goes along with Belyaev's experiment. That owner hasn't seen any changes in aggression, but you would expect to see super-calm dogs eventually develop an uptick in aggression, since fear is a check on aggression and goldens are selectively bred to be low-fear. Aggression is also connected to seizure activity in the brain, and if goldens are starting to develop some seizure-like brain activity (this wouldn't have to be obvious in big, grand mal seizures) you could have aggression.

When you're choosing a mutt, try to pick a dog who comes right up to you and can be friendly. A lot of mutts are horribly distracted inside a kennel or pound, so it can be hard to tell what they'll be like once they've adjusted to a new home, but even at the pound a dog with a good temperament doesn't act terrified.

On the other hand, the Monks of New Skete give different advice. They say that all litters have *loners, aggressors,* and *retreaters.* They say you shouldn't pick the first puppy who comes up to you, because that's the dominant puppy, and he's going to be most prone to having behavior problems. I don't completely agree with that, especially when it comes to mixed breeds. The Monks train German shepherds, so it's possible their observations are more relevant to dogs like shepherds and Rottweilers who've been bred to be guard dogs. If you're choosing a dog from a breed that's naturally nervous or shy, I think you definitely want the most outgoing puppy in the litter.

With all puppies, it's a good idea to give them a quick *startle test.* Clap your hands suddenly, or stomp your feet, and see what the puppy does. All puppies should flinch when they hear a sudden, loud sound, but you don't want a dog who's so terrified that he runs off to the corner of his cage or crate and cowers. Dog trainers use a version of this test to choose puppies who will be good service dogs. They drop a heavy piece of logging chain with four or five links on the floor about four feet away from the puppy. Puppies who get really upset by this aren't the best candidates to work as a service dog for a person with disabilities.

Bone size tells you a lot, too, so look for strong, sturdy bones. You don't have to adopt a hundred-pound monster; just try to find a puppy whose bones aren't tiny and delicate. The same principle applies to horses.

With horses, there's another physical trait you can use in judging a young horse's temperament: the location of his hair whorl. The hair whorl is the round patch of "twisty" hair all cows and horses have up at the top of their faces. The more nervous the animal, the higher the patch. Mark and I were the first to discover this, but trainers have said for a long time that horses with high whorls are more intelligent. What Mark and I realized is that the real difference isn't intelligence, but fear levels. High-fear animals are often smarter, and that's what the trainers picked up on. That was the other thing Mark noticed when he matched up trainers with the kind of horse they were training. Rough trainers had horses with big bones and low hair whorls.

I've already mentioned that although hair color doesn't matter, you want to adopt or buy an animal whose skin isn't too light. I would avoid a puppy that has too many albino characteristics, such as blue eyes, a pink nose, and white fur on most of its body.

Most animals in the wild are either all one color or have an overall mottled, speckled color. Only domestic animals have piebald coloring, where large areas of fur are white. (Zebras and skunks are close, but they probably have too much black fur to be considered piebald.) Belyaev's foxes started out mostly gray and then, as they became domesticated, some of them developed the piebald black-and-white coloring you see in Border collies.

I've been keeping track of animals with white patches of fur, and I've noticed that animals with a white patch of fur someplace on their bodies seem to be less shy than animals without. Ben Kilham, a man who lived with wild bears in the wilderness, actually named one of the bears he knew White Heart because of the patch of white fur on her chest. White Heart was the friendliest bear, the one he could get closest to, and she was the first to be shot by hunters because she didn't have the same fear of people that all black bears did.[29]

Later on I saw a photo of dancing bears in Afghanistan, and every one of those bears had a white fur patch on its chest. I've even started to see this pattern in wildlife photography. Derek Grzelewski, who took a series of photos of otters, mentions that some otters are more "inquisitive" and less "wary" than others.[30] If you look at his pictures of his two inquisitive otters, both of them have white fur at their throats, and one is looking straight at the camera. Those are the only close-ups in the whole batch of photographs, possibly because the solid-colored otters kept their distance.

I don't know whether that tells us anything about what kind of dog a black puppy with a little spot of white on his chest will grow up to be. But I'd be surprised if he was as nutty as some of the Dalmatians out there.

6. How Animals Think

Those pigeons that poop on the cars at the Denver airport can tell the difference between Monet and Picasso. At night they roost in a man-made concrete rookery located over the most expensive parking spots at the airport. When wealthy travelers get back from their trips they find their Land Rovers and Lexuses dribbled over with pigeon poop. For travelers, those birds are a major nuisance, like rats with feathers.

They are also potential art connoisseurs. George Page, in his book *Inside the Animal Mind,* describes a famous experiment in which pigeons were taught to distinguish between paintings by Picasso and paintings by Monet.[1] The birds learned the difference easily. A pigeon can quickly learn to peck at a painting by Picasso, instead of a painting by Monet, and vice versa. Not only that, but when the experimenters showed the birds a painting by *Manet* (not Monet), whose style is similar to early Picasso's, the pigeons pecked the Manet, too. The birds make the same mistake entry-level art students do.[2]

Another experiment showed that pigeons who had never seen a tree in their lives, because they'd been born and raised in a lab, could easily learn to peck at a picture that contained a tree. That might not seem so amazing, except for the fact that they could also peck a picture that contained just one tiny part of a tree. They understood that a part of a tree was still a tree, even though technically a solitary leaf doesn't look anything like a whole tree.[3]

Pigeons are a lot smarter than people think.

Animal researchers are finally beginning to catch up to the little old ladies in tennis shoes who say Fifi the poodle can think. But it's still a battle. The fights are always between a big group of experts who think animals don't have a lot of feelings or aren't very smart, and a much smaller group of researchers who think there's a lot more going on inside an animal's head than we know. The really nasty fights always seem to go one way: it's always the animal "debunkers" who are on the attack. At least, I don't remember a single big academic fight where someone got fired or lost their funding for doing a study where the animal turned out to be *dumber* than people thought, and lots of studies like that have been done. Claiming that an animal *can't* do something isn't considered blasphemous.

Fortunately, it's gotten a lot more respectable to argue that animals are smarter than we realize. One of the main research teams we can thank for that is Dr. Irene Pepperberg and her twenty-five-year-old African gray parrot, Alex. Alex has now reached the cognitive level of a *normal four-to-six-year-old child*.[4]

His achievements are nothing short of revolutionary, because up until Alex came along no one had ever been able to teach birds much of anything at all. It wasn't because they hadn't tried, either. Bird researchers had spent hours and hours trying to teach birds concepts like color, and no bird had even come close to figuring it out. Birds couldn't even learn labels for familiar objects, something everyone agreed apes could do. Even though experts were extremely skeptical of the language abilities of apes like Kanzi, who was said to have *receptive language* equivalent to that of a two-and-a-half-year-old child, it was obvious that you could teach an ape a huge amount. But birds seemed like real birdbrains. (Receptive language means the language you can understand, as opposed to *expressive language*, which is the language you can use to speak or write.)[5]

So it was a huge shock when Dr. Pepperberg succeeded where every single person before her had failed. Not only could Alex learn categories like color and shape, which no bird had ever done before, he learned them easily. Also, once he'd learned the categories, he could spontaneously answer questions like "What color?" and "What shape?" about brand-new objects he'd never seen before.

This means Alex was learning *abstract categories* like color and shape, not just *concrete categories* like "cat" and "dog." Dr. Pepperberg says the difference between concrete categories and abstract categories is the difference between *classification* and *reclassification*. We use simple classification, like sorting out dogs and cats, to form basic, concrete categories. Concrete categories are permanent and stable. A dog is never going to be a cat, and a cat is never going to be a dog.

But when you're using abstract categories to classify things, objects can jump categories. A blue triangle can be grouped with blue squares or with red triangles, depending on which abstract category, color or shape, you're using to make the classification.

A lot of researchers have shown that animals form concrete categories. It would be extremely surprising if they didn't, since an animal has to be able to distinguish between basic categories like food/not food and shelter/not shelter in order to survive.

But the research on whether or not animals can handle the most abstract categories still hasn't produced a firm answer. We know that abstract categories like color are hard for young children to learn. At first, a child will learn that grass and broccoli are green, and apples and roses are red, without figuring out that there's such a thing as *greenness* or *redness* as a separate category unto itself. *Greenness* and *redness* are just part of the apple. Animal behaviorists always assumed that if forming abstract categories is hard for children, it was probably impossible for animals. But now, thanks to Dr. Pepperberg and Alex, we know it's not.

Alex can reclassify objects on demand. If Dr. Pepperberg shows him a square piece of blue wood and asks him, "What color?" he'll say, "Blue." Then if she asks him, "What shape?" he says, "Four-corner." For Alex color and shape are abstract categories that can apply to any object, not just to the objects he's been taught.

DO ANIMALS HAVE TRUE COGNITION?

I like the way Marion Stamp Dawkins, a researcher at Oxford who studies animal behavior and thinking, defines thinking in animals. She starts by saying what *true cognition* is not. True cognition is not hardwired instinctual behavior, and it is not learning a simple rule of thumb.[6]

True cognition, Dr. Dawkins say, happens when an animal *solves a problem under novel conditions.*

By that definition, birds are star performers. One of my favorite bird experiments is the one with about the thieving blue jays. Blue jays are famous food thieves who, in nature, know enough to hide their food so other jays won't get it.

The researchers set up a situation where a jay would have to hide some food in the presence of other jays who were watching him. They gave the first set of blue jays some mealworms and a refrigerator ice tray filled with sand. The jays all hid their worms in the trays, while the other jays watched.

Then the experimenters took the watcher jays away—and the blue jays immediately dug up their mealworms and re-hid them in other parts of the tray. They obviously knew the watcher jays would try to steal their food, and they also knew the other jays knew where they'd *hidden* the food. So they hid the food again.

That is true cognition. The blue jays were in a novel situation, and they figured out a solution.

Mark saw two magpies using a similar strategy on Red Dog. Red Dog was eating a marrow bone that the magpies wanted for themselves. So the birds teamed up to get Red Dog away from the bone. One bird would lure Red Dog into chasing him, and the other would fly down to the marrow bone and start eating it. Then when Red Dog came back to the bone and chased *that* bird away, the first bird would get its turn to eat some of the marrow. The birds were double-teaming Red Dog.

There's been one formal experiment on ravens tricking each other to get food. The researchers studied two ravens, a dominant male and a subordinate male. At the beginning of the experiment the subordinate male found most of the food the experimenters had hidden, and the dominant raven chased him away and took the food for himself. So the subordinate male started tricking the dominant male by heading off to boxes he already knew didn't have food. Then when the dominant male followed him and chased him off, the subordinate male had a head start to the boxes that *did* have food. That worked for a while until the dominant male stopped following him and looked for food on his own.[7]

Crows are really smart birds, too. The Betty and Abel study shocked the world when it appeared in *Science*.[8] In the study the researchers were testing two crows, Betty and Abel, to see whether they would choose a hooked wire or a straight wire to use for getting some food out of a tube. During one session Abel snatched the hooked wire away from Betty, leaving Betty with only the straight wire to use. When she realized the straight wire wouldn't work, she bent it into a hook. She did this nine different times, using different techniques. She also made improvements to her hook after using it, changing the angle to make it just right.

No one had ever seen any animal do anything like this, ever. It wasn't that long ago researchers believed man was the only animal to use tools at all. Then, when people finally discovered chimpanzees using tools in the 1960s and 1970s, no one ever saw them actually *manufacture* a tool. The chimpanzees would just pick up an object in the environment, like a twig or a leaf, and stick it down a termite mound to fish out some termites to eat. Betty's tool creation is even more amazing when you consider the fact that Betty didn't know anything about wire or its properties and didn't have any *reason* to know anything about wire and its properties. In nature nothing bends and holds its shape the way wire does.

I heard another amazing crow story from a man I know. He's fed up with a crow who is damaging his house. I can really relate to that. There's a crow in my neighborhood who has spent the last five years of his life dismantling and pulling out the rubber weather stripping in my bathroom skylight. It's taken him five years to pull out a six-inch strip, and he just keeps at it. He's so dedicated to his project his behavior seems instinct-driven and almost obsessive-compulsive.

I can't get him to stop. I throw hats at the skylight from inside the bathroom to scare him off, but he always comes back. If he keeps doing it the skylight is going to leak, but what I really worry about is that if he finally gets all the weather stripping out he's going to eat it and get sick or die. This is where blind instinct overrides cognition: a bird that's so smart some of the time can be so stupid other times.

The man I know apparently has a similar situation with a crow at his house, only he's opted to use a weapon more dangerous than a soft hat. But he's never been able to shoot his invader, because the crow always knows when he's thinking about getting his gun. The bird will be there in the man's yard attacking the house while the man does his yard work, but the minute the man goes inside the house to get his gun the bird is gone. This has happened over and over again. The homeowner is completely mystified. When he goes inside his house without any intention of getting his gun, the crow stays in the yard. When he goes inside his house with the intention of getting his gun, the crow takes off.

How does the bird know it's time to get out of there? Probably the crow has picked up on differences in the man's behavior. I'm guessing that when the man gets irritated enough to go get his gun, first he does a lot of angry staring at the bird. The crow knows that's dangerous and takes off.

No one has ever seen a dog make a tool, but dogs can problem-solve in novel situations. Guide dogs for blind people have to be able to respond appropriately in new situations. Some service dogs are better problem-solvers than others, of course. In one city, highway engineers wanted to save money on curb cuts for wheelchairs at intersections. Normally a street corner will have eight curb cuts, one on each side of the four corners. To economize, the engineers reduced the number to four, putting each curb cut at the *point* of each corner, facing diagonally across the intersection.

That was a problem for the service dogs, who had all been trained on eight-cut corners. Some dogs got confused by the new design and took their owners clear across the intersection on a diagonal. But the really smart dogs led their owners down the diagonal curb cut and then back around to where the curb cuts would have been located in the normal crosswalk design. *Then* they crossed the street. That's problem solving in a novel situation.

The wild dogs in Mexico City go our service dogs one better. They cross the street in packs, with the light, in the intersection. They probably learned how to do this by watching how people cross the street.

Elizabeth Marshall Thomas, who wrote *The Hidden Life of Dogs*, discovered that her dogs had figured out on their own that intersections are dangerous.[9] To avoid getting hit by turning cars, her free-roaming dogs learned to cross the street in the middle of a block instead of at the intersection. That way they could see all the cars that were coming toward them from a distance, and not be surprised by a car making a sudden right or left turn into their path.

In farming and ranching you see lots of situations where animals will learn something useful by accident, such as how to break through a fence or open a gate. This is probably not true cognition, but some of these animals are pretty clever, and in the field it's hard to say what's true cognition and what's not. Most cattle and horses will never touch gate latches to try to open them, even though they've seen people open the latch a thousand times. However, if an animal accidentally learns to open the gate he'll never stop. He won't *unlearn* it, and he generally can't be trained out of it. My aunt had a horse that learned to put his head through a gate and lift it off the hinges. The only way we could get him to stop was to install a bracket at the top of the fence. Once one animal figures out how to open a gate, the other animals can learn how to open the gate by observation. *Then* you've got a real problem on your hands.

The big problem is fence busting. Every year I get about twenty calls from lawyers about cattle getting loose on the highway and getting hit by a car. The drivers always want to sue the ranchers for inadequate fencing. I have to explain to the lawyers that there is no pasture fence on the market that can keep cattle inside a pasture once the cattle have learned how to get through it. Only a steel stockyard fence is strong enough to hold cattle in physically, and steel fences are too expensive to put up around grazing lands. The fences ranchers use keep cattle in only because the cattle don't realize that they have the power to break through them.

Is fence busting true cognition? Sometimes it is and sometimes it isn't. Usually cattle discover how to break through a fence by accident. Cattle will push on a fence to reach greener grass on the other side, and keep pushing until one day the fence falls over. Then they draw the appropriate conclusion: if I push on the fence, I can get out and go eat where I want. Animals also figure out, probably by accident, that if they run through an electric fence it's going to hurt for only a few seconds. We know this because pigs who have learned to go through electric fences often squeal *before* they hit the wire. They know what's coming.

Some cattle have learned to break through a fence through simple trial and error, but others have started to build on what they learned by accident. There was one bull from the Arizona high country who was the champion fence buster. Bulls are the worst fence busters, and once a bull has learned to break through a fence it's difficult to keep him in. This particular bull was the champion; he took out fences faster than the U.S. Forest Service could build them. He knew how to knock over a high-quality four-strand barbed wire fence built to government standards. In one afternoon he walked through four brand-new fences. I saw him after he had been locked in a stall corral that was too strong for him to break out of.

All of us were amazed that the bull could tear out so many barbed wire fences without getting cut. His tan-and-white hide did not have a single scratch. This is where cognition is at work. He had figured out how to knock over a barbed wire fence without getting cut. Nobody ever saw him do it but he must have figured out that if he pushed over the posts with his head first, and then walked through, he would not get cut. He was careful.

Holstein steers are another story. With all the licking and tongue manipulating they do, Holsteins end up opening gate latches beef cattle never even try to open. I don't think they're really solving a problem, though; it's more like a happy accident. What starts out as a pure desire to lick and tongue things turns into the discovery that they can open gates. Still, once they figure it out they're experts. They can open just about anything, including every sliding bolt gate latch on the market. The only kind of latch that can keep a Holstein cow inside a pen is a chain hooked together with a dog leash snap. They love to get out, too. At one feedlot a group of Holsteins escaped their pen and walked up to the office to lick the windows and remove the paint from the manager's pickup.

ARE ANIMALS AS SMART AS PEOPLE?

I can't answer that question, and neither can anyone else. Researchers who believe we know for a fact that man is the crown of creation when it comes to IQ are off base. That's what researchers *think*, not what they know. I've come to the conclusion that although in many ways other mammals are similar to us, in other ways they may be totally alien. A lot of our tests and experiments with animals probably aren't telling us what we think they're telling us.

Dr. Pepperberg's breakthrough with Alex ought to make researchers think twice. It's not just that what we know keeps changing, but that the way we go about *finding out* how animals think sometimes changes, too. That's the moral of Dr. Pepperberg's story. The reason she finally succeeded where everyone else had failed was that she was the first person to consider that maybe it was the researchers' fault birds weren't learning anything, not the birds'.

All of the parrot studies up to then had used a classical *operant conditioning* format. Operant conditioning, also called *instrumental conditioning* or *stimulus-response teaching,* is when an animal learns *to do something* in order to get what he wants. A rat who's learned to press a lever to get food pellets has had operant conditioning. Using operant conditioning the experimenter would show the bird a red triangle and a blue triangle and say "touch blue," then reward him with a piece of food whenever he happened to peck at the blue triangle by chance. If he happened to peck the red one he didn't get the food. After a while he was supposed to learn blue, because he had been rewarded for pecking the blue triangle every time he heard "touch blue." That's classic behaviorism.

The problem was, no bird ever learned blue. They didn't learn red, either. They didn't learn anything, really. Apes weren't learning too much in those setups, either, but no one wanted to hear about it, because everyone thought it was much more scientific to do a stimulus-response experiment in the lab than to watch an animal learn things naturally in his normal habitat. When a few researchers began teaching apes in more naturalistic settings they were criticized for being unscientific and performing *uncontrolled* experiments. In science, there's nothing worse than an experiment that's uncontrolled.[10]

Dr. Pepperberg decided to give up on operant conditioning and try a different branch of behaviorism called *social modeling theory.* Albert Bandura developed social modeling theory at Stanford University in the 1970s, based on how he thought real people and real animals probably learned in the real world.[11] For years behaviorists had assumed that animals and people learn everything they know through either *operant* or *classical conditioning. (Classical condition-*

ing works with innate, reflexive responses like eye blinks and saliva-tion. Pavlov's dog learning to salivate at the sound of a tone is classi-cal conditioning.)

But Dr. Bandura pointed out that the stimulus-response learning animals did in labs was just learning by trial and error. The animal does more of whatever behaviors he gets rewarded for doing, and less of whatever behaviors he's been punished or negatively reinforced for doing.

That sounds like a logical way to learn until you think what it would mean in the wild. In the real world, trial and error learning would get a lot of animals killed. If the only way a baby antelope could learn to run away from a lion was by finding out what happens if you *don't* run away from a lion, there wouldn't be any baby ante-lope left. Pretty soon there wouldn't be any lions left, either, since they wouldn't have baby antelope to eat.

In Dr. Bandura's view, animals and people had to do a huge amount of observational learning. He thought that a baby antelope would learn to run away from lions by watching other antelope run away from lions and doing the same thing. Today we know Dr. Bandura was right, partly thanks to Susan Mineka's research on monkeys and snakes.

Dr. Bandura had obviously hit on something with social modeling theory, but it didn't occur to anyone to try using it in their research on animal learning. That was Dr. Pepperberg's innovation. She set up a so-cial modeling situation for Alex. Instead of teaching Alex one-on-one she taught him two-on-one, two people to one bird. And instead of teaching Alex directly, she *taught the other person,* while Alex sat on his perch and watched. No one had ever done that before.

She also used items a parrot really, really wants, like a nice, crunchy piece of bark, for her learning materials. Animals and people both pay more attention to things that are important to them, like food, and you have to pay attention to learn. A parrot in the wild doesn't care about blue triangles, so why should he care about blue triangles in the lab? He doesn't.

So if Dr. Pepperberg wanted Alex to learn the color blue, she took a nice, crunchy piece of bark and painted it blue. Then she'd sit down with Alex and her research assistant and ask the assistant, "What color?"

If the assistant got the answer right, he got to play with the bark.

If the assistant got the answer wrong, he didn't get to play with the bark. All Alex got to do was watch. Dr. Pepperberg called her technique *model/rival*, because the assistant was a *model* for Alex to copy and also a *rival* for whatever item Dr. Pepperberg was using in her lesson. She set up a competition for scarce resources between Alex and the assistant.

Using modeling theory was the breakthrough. Alex learned so much that he started asking questions on his own! One day he looked at his reflection in the mirror and asked Dr. Pepperberg, "What color?"

After he'd asked about his own color six different times, and heard answers like "That's gray; you're a gray parrot" six different times, he knew gray as a category. From then on he could tell his trainer whether or not any object she showed him was gray.

This is nothing short of miraculous as far as I'm concerned. Alex was never taught to ask questions; he just did so on his own, spontaneously. That's incredible, because question asking seems to be a separate skill from making statements, judging by the language of autistic children. Autistic children who can talk rarely ask questions; some of them never do. I know a mom whose sixteen-year-old has been talking since the age of two, and she says to this day she can count on one hand the number of questions he has asked.

Question asking is so important that Bob and Lynn Koegel, of the Autism Research and Training Center at the University of California, Santa Barbara, made major breakthroughs in their autism clinic when they started teaching autistic children to ask questions.[12] I wonder whether we would have major breakthroughs in language comprehension with apes and dolphins if we taught them to ask questions, instead of just having them answer questions all the time.

LEARNING THAT'S EASY FOR PEOPLE, HARD FOR ANIMALS

Most birds and animals are almost certainly smarter than we know, but that doesn't mean they don't have some limitations that humans don't. (Humans have limitations animals don't, too. I'll get to that in the next chapter.)

I've said several times now that one of the major differences between people and our fellow mammals is that we have larger, better-developed frontal lobes. One of the benefits of having bigger frontal lobes is that we have more working memory. Since working memory is an important factor in general intelligence, if animals have less working memory overall, that's going to make a difference in their general cognitive abilities.

The question is, what differences are you going to see in a person or animal with lots of working memory versus a person or animal with a lot less working memory? I think my own brain is a good place to start, since I have terrible working memory. If I were a computer I would have a huge hard drive memory and a very small microprocessor. As a result, I have a hard time doing things that involve multitasking, like trying to make change and talk at the same time. Another problem area for me: mental arithmetic. I can't hold one number in memory while I manipulate another. For me to try to add up two two-digit numbers inside my head would be a stretch, and I couldn't even begin to add two three-digit numbers together without writing them down where I can see them.

Since we never ask animals to multitask or add numbers in their heads, one of the main places you can see this difference is in situations that require an animal to be good at *sequencing*. (I'm talking about primates and domestic animals, not birds and sea mammals like dolphins. Birds and dolphins have different brain structures from ours, and I don't know enough about their sequencing abilities to comment.) Animals are not good at sequencing. A good example is dogs getting tangled up in leashes or tie-outs. Owners are always amazed at how helpless a dog is once he's gotten his tie-out wrapped around a tree.

A big part of the problem is that he can't remember the sequence of events that got him to where he is, so he can't retrace his steps. He has the same problem if he just tries to start fresh and figure it out. If one move doesn't work he has to be able to hold that failure in mind while testing other moves. A dog probably doesn't have enough working memory to do that. He's like a person who gets mixed up driving unfamiliar streets after dark. A normal person with an excellent working memory can end up going around in circles in that situation, because he's hit the limits of his working memory. He can't hold all of the different routes he's tried in working memory while he tries new ones, so he keeps going over the same route all over again without realizing it until he ends up back where he started.

Dogs can learn sequences, like the ones working dogs perform at show, with a lot of direct training. However, I think it's probably as hard for a dog to learn show sequences as it was for me to learn the sequence of events that take place in a large meatpacking plant. When I first went into a big plant the place looked so complicated I was amazed the managers were able to keep track of all the complex procedures. I didn't know how anyone could understand and remember anything so intricate.

In the early 1970s I visited a big meatpacking plant every Tuesday afternoon for three years. I used to stand for hours on a catwalk overlooking the floor where the carcasses were processed and dressed by about a hundred employees altogether. The place was a mass of visual details, and every Tuesday afternoon I downloaded more details into my brain.

At first I tuned in to all the really minute details that attracted my attention. Bob, the plant superintendent, was surprised that I kept asking him questions about small details such as how they attached a chain to the hide during hide removal. Apparently nonautistic people could get the gist of the place without having to know every little thing about it. But I couldn't.

One disadvantage of my type of thinking that I probably share with animals is that it takes a long time to download enough details to learn a complex sequence. To do it, I have to create a computer video in my imagination. With the plant, all told, it took six months to download a complete videotape of the entire place into my head. Twenty-four Tuesday afternoons.

Then one day I was standing on the catwalk and suddenly it all seemed simple. I didn't have to worry about remembering the sequence anymore, because I could walk through the whole plant in my mind. Every step in the sequence was connected to the next step, so I didn't have to hold hundreds of different, separate details in my working memory at the same time. I just had to remember one step at a time, and that step brought up the next step.

For me, trying to learn a sequence or add numbers in my head is like having more than one window open on your desktop. If I'm trying to add 49 to 56, first I add 9 and 6 to get 15 and carry the 1. That's in the first window.

But then it takes a really long time for me to close the 9 plus 6 window and open a new window to handle 4 plus 5. By the time the new window is open I no longer remember the 4 and the 5. Or, if I do manage to remember the 4 and the 5 (plus the 1 I have to carry), it takes so long to close the 4 and 5 window and reopen the 9 and 6 window that I've forgotten the original 15. I can work inside only one window at a time, and it takes me forever to switch to a different one. I wonder if animals are like that, too.

The breakthrough with the meat plant came when I could put the whole plant in one window and not have to switch back and forth. Then I could understand and remember it, and after that when I visited other meat plants I could easily pick out the familiar machines even though the floor layout was different. A dog probably has to get any sequence he's learning into one window, too. I suspect that once that happens, the dog "gets it" the way a person "gets it." He understands what he's doing and can apply it to new situations. That's my guess.

THE MAN WITHOUT WORDS

In 1974 the philosopher Thomas Nagel wrote an essay called "What Is It Like to Be a Bat?" that researchers have been arguing about ever since.[13] I think most researchers, thirty years later, would say it's impossible to know what it's like to be a bat, although they disagree with each other about why.

To me, "What is it like to be a bat?" isn't the right question. It's too absolute. I'm never going to know what it's like to be a bat, and a bat's never going to know what it's like to be me. Of course, Professor Nagel wasn't just talking about empathy; he was talking about the scientific method and whether you could ever fully explain consciousness in terms of brain biology. But that doesn't change my point. The fact that it's impossible to know what it's like to be a bat doesn't mean it's impossible to know *anything* about being a bat.

Since almost all researchers believe that animals don't have language, a good place to look for an answer is in the lives of *people* who have no language. We've already seen that autistic people have a lot in common with animals, but another source of clues comes from normal people with normal brains who don't have language. *How do language-less human beings think?*

There are probably lots of language-less people in the world. Usually they are people who were born deaf into communities too small to have anyone who spoke sign language, and too poor to have schools for the deaf. But there are also some language-less people who were born into middle-class American homes but were never taught sign. Their brains are normal, and they had normal parents with normal incomes who loved them. They weren't poor and they weren't abused. The only reason they don't have language is that they were never exposed to language. (Probably in many of these cases the parents believed that allowing their children to learn sign would prevent them from using whatever residual hearing they had.)

The strange thing is that practically no one has studied these people. When I did a Google search for the phrase "language-less people" only nine entries came up. It's bizarre. It's especially strange when you consider how much attention has been paid to feral children and to horribly abused children like Genie, the thirteen-year-old California girl who grew up without language because her father strapped her to a potty chair around the age of twenty months and didn't allow her to have any human interaction. When Genie's mother finally brought her to a welfare office, she had only two words, "stopit" and "nomore."[14] A case like Genie's is extremely interesting, of course, but she was emotionally abused and nutritionally deprived. It's hard to tell how much relevance her cognitive skills have to a normal language-less animal or autistic person's cognitive skills.

Why aren't normal language-less people on the agenda?

The best book on a normal language-less person is *A Man Without Words* by Susan Schaller. Susan Schaller has spent twenty years traveling and researching language-less people completely on her own. The experts she tried to get help from when she first started out were dismissive, uncooperative, or hostile. She even got yelled at by one researcher who shouted, "Who are you?" A graduate student told her, "Nobody's interested in that subject anymore—that was popular last century."[15]

Susan became interested in language-less people when she volunteered to teach Ildefonso, a deaf mute Mexican immigrant who was raised in a town that had no education for deaf children. *A Man Without Words* is the story of her work with him. Susan discovered that Ildefonso had no concept of language at all. Later she learned he had a deaf brother, and that the two of them had figured out some simple ways to communicate as children. But he had absolutely no idea that spoken or written language existed. He understood that the other children did something important with their schoolbooks, but he did not know what it was.

It took Ildefonso only six days with Susan to grasp the idea of language. In the book, he has a revelation that's a lot like the water pump scene in *The Miracle Worker* when Helen Keller suddenly understands what language is.

Although he got the *idea* of language quickly, it took much longer for him to be able to learn and use the language Susan was trying to teach him. One of the most powerful parts of the book, for me, is the day when Susan tries to teach him the words for color. Susan is teaching him the names for colors, like red, yellow, and green, but when they get to "green" suddenly he becomes highly agitated and mimes running and hiding while signing "Green!! Green!!"

Susan couldn't understand why he was so frantic, until she learned that green was the most important concept in Ildefonso's life. Ildefonso was an illegal immigrant who supported himself working harvesting crops and picking apples. All the good things in life and all the bad things in life were green. Green money and picking green crops let him feed his family in Mexico. Border Patrol agents wearing green uniforms and driving green trucks were the bad people who would grab him and take him back to Mexico, to the place where there was less work and food was scarce.

The most important thing in life was the Green Card that magically repelled the bad green men.

Susan writes that it was impossible for her to imagine Ildefonso's world. I expect she knows a lot more about the world of language-less people now that she's spent two decades searching them out, and I'm looking forward to her next book. She did perceive differences in Ildefonso that I think directly apply to animals, as well as to people with autism.

The main difference between Ildefonso and people who have language is that he was missing a layer of abstract thinking. For instance, he didn't have the categories of *real* and *fake*. He just knew that some Green Cards worked to keep the green men from taking you back to Mexico, and some Green Cards didn't. He didn't know why.

He also didn't have *just* and *unjust* as abstract categories. It's not that he didn't have morals or a conscience. Susan doesn't say a lot about this, but she writes that Ildefonso became upset one day when she kept insisting on paying for his lunch after he had signed that he wanted to pay. Ildefonso got more and more angry until finally he signed, "God. Friend. Burrito buy I."

"He connected *God* and *friend* and placed them above burrito buying," Susan writes. "His anger was that of a religious instructor. I was properly rebuked for my concern for the material world. Who had more money was trivial." Later on he asked her what "God" meant, but he had already figured it out on his own. Susan writes that he had guessed that the word "God" stood for "unseen greatness, apart from and more important than the tangible stuff in front of us."

Although Ildefonso had the idea that there was something greater than the material world, he didn't seem to have any concept of human justice. He had no idea whether it was just or unjust for the green men to catch him and take him back to Mexico; he just knew that's what the green men did, so he needed to stay away from the green men. He was trying to understand the rules, without realizing there were principles *behind* the rules.

Ildefonso was an innocent. He didn't see all the good and bad that people do, and he didn't know there could be good and bad rules, either. After he learned language, he was sad to learn of the terrible things people do. Animals are innocents, too. Even when animals are treated badly by humans, or see other animals treated badly by humans, they don't seem to develop the abstract categories of just and unjust. Like Ildefonso, animals try to learn the rules without seeming to realize there are principles behind the rules. Since they don't know there are principles underlying the rules they don't realize that the rule itself can be just or unjust, or that a person could be breaking abstract principles of justice. Animals live much closer to the plain facts of the situation.

But the important thing to realize is that Ildefonso's innocence was not the same thing as being stupid, or unable to think. Ildefonso wasn't stupid, and he functioned as a person of normal intelligence and reasoning ability or even above-average intelligence, given that he had been able to immigrate to a foreign country, find work, and manage his life while struggling with a huge disability.

This means that when it comes to animals, we should not equate innocence with lack of intelligence. The fact that a dog never rejects a nasty owner doesn't make him stupid. It makes him innocent. Dogs may well have lower reasoning ability and general intelligence than people do, but a dog's "blind devotion" isn't evidence one way or another.

Although Ildefonso didn't have an abstract sense of just and unjust, he did have an immediate, concrete sense of right and wrong, which he showed when he gave Susan the stern lecture on friendship. That shows that you don't have to have language to have a conscience, which means it's at least *possible* for an animal to have a conscience, too. Many owners have seen their dogs act remorseful after doing something wrong, but animal behaviorists always reject this interpretation. However, no one has shown that an innocent animal *can't* feel bad for doing something he knows is wrong, the same way an innocent child can feel bad for doing something he knows is wrong. We shouldn't assume that we know for a fact animals never experience the emotion of guilt, because we don't.

A friend of mine has a story about one of her dogs showing remorse that I think is probably right. She has two dogs, a male and a slightly younger female, and she had taken them for a walk with one dog on-leash and one dog off-leash. Unfortunately, when she got up the hill close to her house a neighbor saw them and started yelling at her about the loose dog.

Since she didn't have another leash with her, she had to thread the leash through one dog's collar and hook it to the other dog's collar, which meant their heads were pulled so close together they were touching. The dominant dog didn't like that at all, because dominant dogs guard their body space closely and need more of it. So this was a violation of his dominant-dog rule.

They had to walk all the way home like that, with the dominant dog looking more and more irritated and tense. Finally, when they got back to their own driveway, the dominant dog snapped. He burst out in a loud snarl and bit his housemate on the nose, something he had never done before. The younger dog shrieked.

My friend jumped over to the dogs and got them unhooked, but the dominant dog didn't run off to his freedom. He stayed right by the subordinate dog, licking and licking her on the lips. My friend said he looked mortified. She'd never seen him kiss his pack mate like that, and it was obvious to her, as well as to her next-door neighbor, who saw the whole thing, that he was sorry for what he'd done and was trying to make it up to his friend. He acted like he felt remorse, and I don't think you can rule it out. He was the alpha, and he didn't need to be kissing the subordinate dog to keep on being the alpha. If anything, it was the subordinate dog who should have been doing the groveling, not the dominant dog. But she didn't. She accepted his kisses, and they went back to being friends.

Even though Ildefonso was an innocent, a lot of the abstract "reality" people express through language was still there. Religion is a good example. Ildefonso had gone to church when he was little, but he didn't know what any of it meant, although he instantly figured out that the baby Jesus in a crèche in his adult classroom was the same as the grown-up Jesus he had seen on crucifixes, which I think is pretty amazing.

Although he didn't know anything about the Christian religion his family practiced, he still had a religious sense. This is obvious to me from the fact that he picked up the word "God" within three weeks of first discovering language, and understood that "God" meant "unseen greatness."

I think some of the other language-less Mexicans Susan met years later probably also had a religious sense. She says that Ildefonso's language-less friends, some of whom were living together, treated their precious collection of Green Cards like they were "gold." To me it sounded as if they were treating them like magic, not gold. They had a special place in their house where they kept the cards, like a shrine. The cards were like a religious idol or talisman that could protect them from the evil green men, and their "religion" was like the pagan religions indigenous populations have. The cards were also their savior, the way to get into the Promised Land where there was more food and jobs.

The men didn't know the difference between valid Green Cards and fake ones, and they probably didn't even have the abstract categories for *fake* and *real*. But over time they would have realized that some of the Green Cards had more magic than others, because if you get caught by a green man and you show him one of the Green Cards, sometimes it works but other times he takes it away from you. So some cards are more powerful than others. In religion, you don't test God; you don't stand in front of a train and say God should save you. That's how the language-less men would have felt about the cards. You don't test the cards; it's not right. So you stay away from the green men.

Religion is probably hardwired into the human brain, so it doesn't surprise me that a religious feeling or sense managed to shine through in Ildefonso even without words.[16] By the same token, it wouldn't surprise me if animals have religious feelings like Ildefonso's or a sense of some higher reality or unseen world they can't express. Do some animals have religious feelings and perceptions? Do animals believe in magic? I don't think anyone can rule it out.

The lesson from Ildefonso is that although language does make thought more abstract, without language you can think more abstract thoughts than probably anyone has believed possible. Dr. Pepperberg says the real question about language and animals should be: *at what point* do concepts get so complex that you *have* to have language to form them?

WORDS GET IN THE WAY

One aspect of Ildefonso's mental life that Susan Schaller doesn't mention is his memory for visual detail. I wonder whether his visual memory was superior to a normal person's visual memory before he learned sign. Research shows that language suppresses visual memory. This is called *verbal overshadowing* and is a well-established phenomenon, which I mentioned in Chapter 3. For example, in one study people watched a short videotape of a bank robbery, then spent twenty minutes doing something unrelated.[17] Then one group spent five minutes writing down everything they could remember about the bank robber's face, while the other group did an unrelated task.

Two thirds of the people who wrote nothing down and did unrelated tasks could identify a photograph of the robber, while only *one third* of the people who wrote verbal descriptions could pick him out. This is a well-established effect; many studies have found exactly the same thing, and some studies have extended the effect to auditory memory as well. People who write down a description of a voice are less able to pick it out from other voices than people who didn't describe the voice in words.

These studies have also found that language doesn't erase visual memories for good; it just suppresses them. When the researchers asked the people who wrote descriptions to do something nonverbal for a while, like work a puzzle or listen to music, their visual memories came back, and they could identify the bank robber's face as well as the people who hadn't written descriptions in the first place.

I think for normal people language is probably a kind of filter. One of the biggest challenges for an animal or an autistic person is dealing with the barrage of details from the environment. Normal people with language don't have to see all those details consciously. But I see them, and animals do, too. The details never go away, either. If I think of the word "bowl," I instantly see many different bowls in my imagination, such as a ceramic bowl on my desk, a soup bowl at a restaurant I ate at last Sunday, my aunt's salad bowl with her cat sleeping in it, and the Super Bowl football game.

I think that probably happens to animals, too, and I wonder what Ildefonso's visual memory was like while he was still a language-less person.

AWAKE AND AWARE—ANIMALS ON THE INSIDE

One last thing about Ildefonso: there's no question he was conscious. Many people over the years have argued that if you don't have language you don't have consciousness. I remember in college when one of my professors told the class that animals weren't conscious because they didn't have words to think in. Since I didn't think in words myself, I was shocked when he said that. If an animal isn't conscious, I remember saying to myself, then I'm going to have to assume I'm not conscious, either.

Obviously I *am* conscious, even though I don't think in words, so there's nothing to say an animal can't be conscious just because an animal doesn't think in words. Ildefonso was conscious, and he had no language at all.

I think animals are conscious, too. My question is: does the horse who's scared to death of black hats see mental images of whatever happened to him over and over again inside his head the same way a person with post-traumatic stress syndrome does? Do animals see pictures of food when they're hungry the way I do? Do they see a picture of water when they're thirsty?

Another question I have is: do animals have constant mental activity the way people do, or do they walk around with their minds a blank?

We know they have constant mental activity of some kind, because their EEGs aren't that different from ours. I expect the content of their consciousness is mostly pictures and probably sounds, too. Animals might even have conscious "thoughts" of smells, touch, or taste.

A report in January 2001 about dreaming mice gives us pretty good indirect evidence that animals think in pictures.[18] In that study, two researchers, Matthew Wilson, a professor of brain and cognitive sciences in the biology department at MIT, and Kenway Louie, a biology graduate student there, implanted electrodes in the brains of mice, then taught them to run a maze. When they recorded all of the mice's neural firings they found the brain wave patterns were so precise they could see in the recordings exactly what a mouse was doing at any given moment: making the first turn left, making the first turn right, running down the first passageway or the second passageway, and so on.

Later on, when the mice were asleep and had gone into the REM phase, Wilson and Louie recorded the exact same firing pattern the mice had shown when they were awake and running the maze. The sleep firings were so exact the researchers could tell where in the maze each mouse was, at any given moment, in his dreams. Since people dream in pictures during REM sleep, this is pretty good evidence that animals dream in images, too. There's no way to know for sure that Wilson and Louie's dreaming mice were seeing images of the maze, since the only way to know what pictures anyone is seeing in a dream is to wake him up and ask him. Of course you can't do that with a mouse. But the fact that the dreaming mice were firing the exact same sequence their brains fired when their eyes were open is a good reason to suspect that mice, like people, see pictures when they dream. It's not a huge leap to assume they probably think in pictures when they're awake.

ANIMAL SPECIALISTS

Animals are probably *cognitive specialists*. Some animals, like the Holsteins, are manipulation specialists. Dogs are smell experts. Other animals, like pigeons, are visual specialists.

Some bird and mammal species that have to remember where they've hidden their food are memory specialists and have extra large brain areas devoted to visual memory. The Clark's nutcracker, a type of crow, buries as many as thirty thousand pine seeds in the fall in a two-hundred-square-mile area, then finds over 90 percent of them during the winter.

Compared to animals and people with autism, normal humans are *generalists*. Typical people are usually good at some things and bad at others, but a person who's really smart in one subject tends to be really smart in a lot of subjects. There is one exception to this, which is that gifted children have greater variability on their IQ test sub-scores than children with normal intelligence. But it's not a case of gifted children being brilliant at some tasks and hopeless at others. They're still highly intelligent overall, and they do extremely well on all the different IQ tests, not just on one or two of them.

One important piece of evidence that may support the idea that people are generalists while animals are specialists is the new findings on *g* or *general intelligence* (also called *general fluid intelligence*). The idea of general intelligence, which is what classic IQ tests measure, has been controversial. People like the psychologist Howard Gardner emphasized *multiple intelligences* over one single, general intelligence, and some psychologists have rejected the idea of *g* altogether.

But new brain research supports the idea that *g* exists, and that it's localized to one spot in the brain: the lateral prefrontal cortex, an area at the top of your head and off to the side that handles working memory, abstract thought, and *response inhibition,* which is stopping yourself from doing something you're on track to do, like answering the phone when it rings. If you've decided not to answer the phone while you fix dinner, your lateral prefrontal cortex has to block the impulse to pick up the receiver every time you hear the phone ring.

Jeremy Gray from Washington University, one of the researchers on the *g* study, found that the higher a subject's general intelligence, the higher the activity level in his lateral prefrontal cortex. Dr. Gray told the *New York Times* that the hardest IQ tasks he used in the experiment are like "trying to remember a new 10-digit telephone number while listening to people who are having an interesting conversation."[19]

This finding fits in with behavioral research showing that being able to integrate a lot of information is a big part of being "school smart." Jennifer Symon, a graduate student working with Bob and Lynn Koegel, did a really interesting study comparing "regular" schoolchildren to children whose teachers said they were gifted.[20] She found that the gifted kids were much better at doing multifaceted tasks. She tested the children using tasks that increased in complexity. In the *one-component task* a child was given four bears that were identical in every respect except color, and asked to choose the blue bear. To do this the child had to pay attention to only one element of the task, color. In the *two-component task* children chose among items that varied along two dimensions, like bears and dogs in different colors. The researcher would ask them to pick the green dog. A typical *three-component task* asked the child to pick the "big polka dot circle," and a *four-component task* asked them to pick two objects, such as the big teddy bear and the little square, that added up to four components altogether.

Dr. Symon found that by age three the gifted children could do *four-component tasks*—tasks where they had to pay attention to and pull together four different things in order to succeed. The regular kids couldn't do four-component tasks until they were six.

No one has tried to find g in an animal or bird brain yet, and I don't know what we'll see when they do. For now, since animals— especially domestic animals—as a group have a smaller and weaker prefrontal cortex, I'm assuming they probably have weaker general intelligence. That probably opens the door for them to become super-specialists, which I'll talk about in the next chapter. For now I'll just say that I think the kind of specialization I see in animals and in autistic people probably depends on having a weaker prefrontal cortex.

THAT'S MY STORY AND I'M STICKING TO IT

There are definitely times when normal people's high level of general intelligence makes them too smart for their own good. My favorite example is the rats who beat the humans in a lever-pressing task. Years ago someone decided to compare rats to humans in the kind of standard operant conditioning task experimenters usually do only with animals. (Remember, operant conditioning means the animal or person gets a reward when he does what the experimenter wants him to do.) The rats and the humans had to look at a TV screen and press the lever anytime a dot appeared in the top half of the screen. The experimenter didn't tell the human subjects that's what they were supposed to do; they had to figure it out for themselves the same way the rats did.

The experiment was set up so that 70 percent of the time the dot was in the top of the screen. Since there wasn't any punishment for a wrong response, the smartest strategy was just to push the bar 100 percent of the time. That way you'd end up getting a reward 70 percent of the time, even though you didn't have a clue what the pattern was.

That's what the rats did. They just kept pressing the bar every time the screen changed.

But the humans never figured this out. They kept trying to come up with a rule, so sometimes they'd press the bar and sometimes they wouldn't, trying to figure it out. Some of them thought they *had* come up with a rule, which they then used to tell them when to press the bar and when not to press the bar. But they were deluded. They hadn't come up with the rule at all, and the rats ended up with lots more rewards than the humans.

I believe the rats did better than the humans either because of weaker frontal lobes or because rats don't have language or both. One thing we do know about humans is that the left brain, which is the conscious language part of the brain, always makes up a story to explain what's going on. Normal people have an *interpreter* in their left brain that takes all the random, contradictory details of whatever they're doing or remembering at the moment, and smoothes everything out into one coherent story. If there are details that don't fit, a lot of times they get edited out or revised. Some left brain stories can be so far off from reality that they sound like confabulations.

The interpreter probably got in the way on the lever-pressing experiment. The human subjects kept trying to come up with a story about the dots, and when they did come up with a story they stuck to it. Then the dot story kept them from realizing they should just forget about the dots and press the lever every time the screen changed.

ANIMAL WELFARE: TAKING CARE OF ANIMALS THE WRONG WAY

Working in animal welfare, I constantly have to reason with normal humans who are too smart for their own good.

My most important contribution to the field has been to take the idea behind Hazard Analysis Critical Control Point analysis, or HACCP (pronounced *hassip*), and apply it to the field of animal welfare. The animal welfare audit I created for U.S. Department of Agriculture is a HACCP-type audit.

My HACCP system works by analyzing the *critical control points* in a farm animal's well-being. I define a critical control point as a *single measurable element that covers a multitude of sins.* For instance, when I'm auditing the animals on a farm, one thing I want to know is whether the animals' legs are sound. There are a lot of things that can affect a cow's ability to walk: bad genes, poor flooring, too much grain in the feed, foot rot, poor hoof care, and rough treatment of the animals. Some regulators will try to measure all of these things, because they think a good audit is a thorough audit.

But that's not my approach. I measure one thing only: *how many cattle are limping?* That's all I need to know, just how many cattle are limping. That one measurement covers the multitude of sins that can cause cattle to go lame. If too many animals are limping, the farm fails the audit and that's it. The only way the farm can pass the next audit is to fix whatever it is that's making their animals lame. If management knows what the problem is, they can get busy fixing it. If they don't know what the problem is, they have to hire someone who can tell them, and *then* fix it.

For my animal welfare audit, I came up with five key measurements inspectors need to take to ensure animals receive humane treatment at a meatpacking plant:

- Percentage of animals stunned, or killed, correctly on the first attempt (this has to be at least 95 percent of the animals).
- Percentage of animals who remain unconscious after stunning (*this must be 100 percent*).
- Percentage of animals who vocalize (squeal, bellow, or moo, meaning "ouch!" or "you're scaring me!") during handling and stunning. Handling includes walking through the alleys and being held in the restraining device for stunning (no more than 3 cattle out of 100).

- Percentage of animals who fall down (animals are terrified of falling down, and this should be no more than 1 out of 100, which is still more than would fall down under good conditions, since animals never fall down if the floor is sound and dry).
- Electric prod usage (no more than 25 percent of the animals).

I also have a list of five acts of abuse that are an automatic failure:

- Dragging a live animal with a chain.
- Running cattle on top of each other on purpose.
- Sticking prods and other objects into sensitive parts of animals.
- Slamming gates on animals on purpose.
- Losing control and beating an animal.

This is all you need to know to rate animal welfare at a meatpacking plant. Just these ten details. You don't need to know if the floor is slippery, something regulators always want to measure. For some reason whenever you start talking about auditing the plants everybody turns into an expert on flooring. I don't need to know anything about the flooring. I just need to know if any of the cattle fell down. If cattle are falling down, there's a problem with the floor, and the plant fails the audit. It's that simple.

The plants love it, *because they can do it*. The audit is totally based on things an auditor can directly observe that have objective outcomes. A steer either moos during handling or he does not.

Another important feature of my audit: people can remember two sets of five items. That level of detail is what normal working memory is built to hold on to.[21]

But I find that people in academia and often in government just don't get it. Most language-based thinkers find it difficult to believe that such a simple audit really works. They're like the people in the lever-pressing experiments; they think simple means wrong. They don't see that each one of the five critical control points measures anywhere from three to ten others that all result in the same bad outcome for the animals.

When highly verbal people get control of the audit process, they tend to make five critical mistakes:

- They write verbal auditing standards that are too subjective and vague, with requirements like "minimal use of electric prod" and "non-slip flooring." Individual inspectors have to figure out for themselves what "minimal use" means. A good audit checklist has objective standards that anyone can see have or have not been met.
- For some reason, highly verbal people have a tendency to measure *inputs,* such as maintenance schedules, employee training records, and equipment design problems, instead of *outputs,* which is how the animals are actually doing. A good animal welfare audit has to measure the animals, not the plant.
- Highly verbal people almost always want to make the audit way too complicated. A 100-item checklist doesn't work nearly as well as a 10-item checklist, and I can prove it.
- Verbal people drift into *paper audits,* in which they audit a plant's records instead of its animals. *A good animal welfare audit has to audit the animals, not the paper and not the plant.*
- Verbal people tend to lose sight of what's important and end up treating small problems the same way they treat big problems.

All five of these mistakes hurt the animals. When you make the audit process more complicated, the auditors veer off into all the fine detail that goes into making a humane slaughterhouse, which leads to wanting to micromanage the plants. Instead of looking at outcomes to the animals, they want to tell the plant how to build its floors. Then they want to send auditors out to inspect the construction to make sure the floors are right. The animal gets lost in the confusion. I don't care about floors. I care about cows. Are they falling down? That's all I need to know.

The other thing that happens is that auditors lose track of what's important. If you give an auditor a 100-item checklist, he'll tend to treat 50 of the items as if they're major, whereas maybe only 10 items are so critical that if the plant fails any one of those 10 it should fail the audit, period. When a plant fails 1 critical item out of 10, it's easy to fail the whole plant. But when it fails that same item on a list of 100, it doesn't look so bad.

Even worse, an auditor working with a long, overly complicated checklist can miss the huge problems completely, even though they're on the list. A friend of mine told me a horrible story about a plant where the stunning equipment wasn't working right, and they had live animals hanging from hooks going down the slaughter line. The USDA inspector missed it. He got focused on some worker who was whacking the pigs too hard on the butt, and he wrote them up for that. Meanwhile the plant had a hideous, enormous problem of live animals on the slaughter line that ought to mean an automatic fail. The inspector didn't see it, or maybe he did see it but it didn't *register* on him.

I think this kind of blindness must have to do with the limits on normal human perception. Somehow, when an inspector has to audit 100 different aspects of a plant's functioning, he stops seeing the lady in the gorilla suit. I'm not saying it's okay to be whacking the pigs, of course. It's not, and it should be corrected. But when the audit checklist is too long, auditors start hyper-focusing on small details and missing the great big details that matter the most.

I've seen this happen many times. About a year ago I visited plants in Europe, where the plants and the inspectors were supposed to continuously monitor and improve 100 different items on a checklist. The plants were horrible.

Sometimes the standards that verbal thinkers want to include aren't even connected to reality. For instance, I've been working with KFC—Kentucky Fried Chicken—to raise standards for animal welfare in the poultry industry, and one of the standards an abstract, verbal thinker will want to put on the audit form is that the lights have to be off for at least four hours every night. Well, how am I going to get out to the farm at 3:00 A.M. to make sure the lights are off? I'm not. And I'm not going to trust the paperwork.

What I need to audit isn't the lights, it's the *outcome* of turning off the lights to the chickens' welfare. The lights have to be off because darkness slows down a baby chick's growth. Today's poultry chicken has been bred to grow so rapidly that its legs can collapse under the weight of its ballooning body. It's awful. Darkness slows down the baby chick's growth just enough to prevent this from happening, so getting those lights off is important, because lameness is a severe problem in chicken welfare. I've been to farms where half of the chickens are lame. When I audit a chicken farm, what I want to know is, can the chickens on this farm walk? If the chickens are lame, something is wrong, and the farm fails the audit.

And I strongly object to paper audits, because anyone can change his paperwork if he wants to. However, a plant can't falsify things I can directly observe. I don't want to see the maintenance records on the stunner. If the stunner is well maintained, it's going to work. That's all I need to know. I've measured broken wings on chickens. I want to see the animals.

The other dangerous thing about paper audits and 100-item checklists is that they can set you up for a situation of things slowly getting worse without anyone knowing it. When you drift away from the animals themselves and start auditing the paperwork, the bad can become normal pretty quickly.

I want to stress this point. Maintaining animal welfare standards in a meatpacking plant is an ongoing responsibility. *The whole principle of HACCP is that you have to keep measuring standards and compliance or everything goes bad on you.* It's kind of like maintaining your weight: you have to keep on top of it. Paper audits end up masking small, incremental declines in standards that result in very large drops in animal welfare.

Unfortunately, to an abstract verbal thinker, a list with 100 different animal welfare items sounds more caring than a list with only 5. But I can prove beyond question that animals in plants undergoing 10-question audits are handled much more humanely than animals in plants undergoing 100-question audits. And it's not just that plants using my checklist do well on the big details. They also do better on the smaller details, because *the smaller details are part of the big ones.*

Even though my list contains only five critical control points, it is so strict that most plants thought they wouldn't be able to pass. But then McDonald's started auditing the plants. In 1999 they threw a major plant off the approved supplier list for flunking the audit, and they suspended some other plants. After that the industry got religion, and boy has the cattle handling changed. Let me tell you, you go out there now and they're handling the cattle *nice.* All of the

272 Animals in Translation

plants being audited using my list treat their animals better than plants using 100-item checklists.

Most large plants are now audited by restaurant chains like McDonald's, Burger King, and Wendy's International. Just four years after McDonald's began requiring its suppliers to audit their plants according to my standards, almost every plant is passing easily. Now when you go into a plant it's like a magical change. I think of all the years up to 1999 as the pre-McDonald's era and the years since then as the post-McDonald's era. Up until 1999 the plants might buy the best equipment, but they didn't manage it. They'd let stuff break, and they didn't spend enough time and money training and supervising their staff or firing people who needed firing. Then as soon as McDonald's started auditing, they were hitting up my Web page to learn the stuff they had to do. There were light-years of change.

For the twenty-five years up to 1999 I'd been putting equipment into plants. Some of the plants used it right, but others just tore it up and ruined it. Now my equipment is perfectly maintained and there's nothing broken on it anywhere. For the first twenty-five years of my career I was a hardware engineer; now finally I'm installing the management software. Training those auditors: that's the software installation for the hardware I put into half the plants in North America.

—

My simple five-point checklist works beautifully. But even though it works, and even though I can show that animals being audited by 100-point checklists are being handled poorly, I have to fight constantly to keep it in place.

DO ANIMALS TALK TO EACH OTHER THE WAY PEOPLE DO?

Those are fighting words in the fields of animal and linguistic research. A lot of people are emotionally invested in the idea that language is the one thing that makes human beings unique. Language is sacrosanct. It's the last boundary standing between man and beast.

Now even this final boundary is being challenged. Con Slobod-chikoff at Northern Arizona University has done some of the most amazing studies in animal communication and cognition.[22] Using sonograms to analyze the distress calls of Gunnison's prairie dog, one of five species of prairie dogs found in the U.S. and Mexico, he has found that prairie dog colonies have a communication system that includes nouns, verbs, and adjectives. They can tell one another what kind of predator is approaching—man, hawk, coyote, dog (noun)—and they can tell each other how fast it's moving (verb). They can say whether a human is carrying a gun or not.

They can also identify individual coyotes and tell one another which one is coming. They can tell the other prairie dogs that the approaching coyote is the one who likes to walk straight through the colony and then suddenly lunge at a prairie dog who's gotten too far away from the entrance to his burrow, or the one who likes to lie patiently by the side of a hole for an hour and wait for his dinner to appear. If the prairie dogs are signaling the approach of a person, they can tell one another something about what color clothing the person is wearing, as well as something about his size and shape (adjectives). They also have a lot of other calls that have not been deciphered.

Dr. Slobodchikoff was able to interpret the calls by videotaping everything, analyzing the sound spectrum, and then watching the video to see what the prairie dog making a distress call was reacting to when he made it. He also watched to see how the other prairie dogs responded. That was an important clue, because he found that the prairie dogs reacted differently to different warnings. If the warning was about a hawk making a dive, all the prairie dogs raced to their burrows and vanished down into holes. But if the hawk was circling overhead, the prairie dogs stopped foraging, stood up in an alert posture, and waited to see what happened next. If the call warned about a human, the prairie dogs all ran for their burrows no matter how fast the human was coming.

Dr. Slobodchikoff also found evidence that prairie dogs aren't born knowing the calls, the way a baby is born knowing how to cry. They have to learn them. He bases this on the fact that the different prairie dog colonies around Flagstaff all have different dialects. Since genetically these animals are almost identical, Dr. Slobodchikoff argues that genetic differences can't explain the differences in the calls. That means the calls have been created by the individual colonies and passed on from one generation to the next.

Is this "real" language? A philosopher of language might say no, but the case against animal language is getting weaker. Different linguists have somewhat different definitions of language, but everyone agrees that language has to have *meaning, productivity* (you can use the same words to make an infinite number of new communications), and *displacement* (you can use language to talk about things that aren't present).

Prairie dogs use their language to refer to real dangers in the real world, so it definitely has meaning.

Their language probably has productivity, too, since they can apply the same adjectives to different animals. Dr. Slobodchikoff has also done some interesting experiments to see what calls prairie dogs would make to an object they'd never seen before.

He built three plywood silhouettes, a skunk, a coyote, and a black oval, and dragged them through the prairie dog colony on a pulley. The prairie dogs gave alarm calls to all three objects, and each prairie dog used the same call for the same plywood object. These calls weren't invented on the spot, either. At least one of the calls—for the plywood coyote—was a variant of an old call Dr. Slobodchikoff had already recorded them using. That's more evidence the prairie dogs were combining their old "words" to describe something new.

Another interesting finding: all three plywood objects were new to the prairie dogs, but the prairie dogs used different calls to identify each one. Dr. Slobodchikoff says that means it's unlikely the prairie dogs were simply using a rote call meaning "something new is coming." He also says that the prairie dogs seem to be using *transformational rules* to create their calls. In human language, a transformational rule allows you to turn words into sentences that make sense. The person listening to you uses the same rules to decode what you're saying. The prairie dogs seem to have a transformational rule based on speed. Depending on how fast a predator is moving, they speed up their calls or slow them down.

We don't know yet whether the prairie dogs ever use their calls to talk about things that aren't present. But since other animals have used language to talk about things that aren't present, there's no reason to assume prairie dogs can't do it, too. Some of the apes whom researchers have trained in English over the years have used their words to talk about food that was in another room and not visible, which is spatial displacement, and at least two of them have used signs to ask about animal companions who had been taken away from them to go to the vet. I think it's unlikely that Dr. Slobodchikoff's prairie dogs would have nouns, adjectives, verbs, semanticity, and productivity without also being able to use their calls to communicate about something that is not immediately present.

WHY PRAIRIE DOGS?

From what we know now, it seems prairie dogs' ability to communicate may be greater than that of animals with more complex brains, including the primates. Why would prairie dogs develop more complex calls than the monkeys? Maybe because they had to. Prairie dogs are *super-prey*—there's almost no meat eater in the vicinity of prairie dog burrows that *doesn't* eat them. Dr. Slobodchikoff's list of prairie dog predators is so long it has animals on it most people have never even heard of: "coyotes, foxes, badgers, golden eagles, red-tailed hawks, ferruginous hawks, harriers [a kind of hawk], black-footed ferrets, domestic dogs, domestic cats, rattlesnakes, and gopher snakes."[23] For eight hundred years Native Americans hunted prairie dogs for food, and today humans hunt them for target practice and sport.

To make things worse, prairie dogs live in the same burrows for hundreds of years. That means every single predator in the vicinity knows exactly where to find them. It also means the prairie dogs get to know the local predators on an individual basis. All told, it's likely prairie dogs are so vulnerable they had to develop a really good system of communication to survive as a species. Dr. Slobodchikoff speculates that instead of looking for animal language in our closest genetic relatives, the primates, we should look at animals with the greatest need for language in order to stay alive.

If he's right, that's probably another blow to the idea that human language is unique. If language naturally evolves to serve the needs of tiny rodents with tiny rodent brains, then what's unique about language isn't the brilliant humans who invented it to communicate high-level abstract thoughts. What's unique about language is that the creatures who develop it are highly vulnerable to being eaten.

THE MUSIC LANGUAGE

I think it's likely that the language of the prairie dogs is a musical language. Dr. Slobodchikoff used special computer programs to analyze the prairie dog calls and found that the calls had different frequency ratios, which he thinks are patterns the prairie dogs created. He theorizes that frequency ratios may form patterns. To put it in simpler language, the calls are different pieces of music.

Sophie Yin at the University of California, Davis, found something similar in analyzing thousands of dog barks. Her analysis shows that dogs have different barks depending on the circumstances.[24] When a dog spots a stranger its barks are rapid and urgent. When a dog is playing, its barks are slower and richer in harmony. No one knows what those harmonies mean, but the fact that they vary consistently depending on the dog's situation tells me they likely have meaning to another dog. Dogs are also highly sensitive to tone of voice, which is the musical part of language.

Some scientists such as Steven Pinker, the cognitive psychologist at MIT who wrote the books *The Language Instinct* and *How the Mind Works*, think music is just evolutionary baggage with no real purpose, but so many birds and animals create music that it doesn't make sense to me that music could simply be so much evolutionary baggage.[25] And if music is just evolutionary baggage, then why does the brain have different areas to analyze the five different components of music? Studies of patients with brain damage have shown that the five distinct brain-processing systems for music are melody, rhythm, meter, tonality, and timbre. My hypothesis is that music is the language of many animals.

Brain scan studies are beginning to offer some support for this idea. A study reported in *Nature Neuroscience* found that the same brain area that understands spoken language—*Broca's area*—also understands music. That's a big finding, because cognitive scientists have always believed that Broca's area handles language and nothing else. So far researchers seem to be interpreting the new findings as possibly meaning that Broca's area may be specific not to language but to processing the "implicit rules that organize complex information, such as music and language" instead.[26]

But I think the explanation could be that cognitive scientists were right in the first place. Maybe Broca's area *does* handle language exclusively, and maybe that's why it also handles music, because music is a language, too—or it could be. It's possible that music, or something like it, once was the human language, and maybe it still is the language of birds and animals.

One thing that makes me believe this is high-functioning autistic people who've told me that when they were children echoing sentences they'd heard on TV, they didn't know that the meaning was in the words. They thought all the meaning was in the *tone*. I can relate to that, because tone of voice is the only social cue I pick up easily. I also know of at least one parent who could communicate with her autistic daughter only through singing. If the mom sang, "Set the table now," her daughter understood. If the mom *said*, "Set the table now," her daughter didn't understand. She got the meaning through the music. I wonder whether this is a case of autistic people falling back on earlier, animal forms of communication that are closer to music.

Probably all parents communicate with babies through music. Sandra Trehub at the University of Toronto points out that lullabies are found in every culture, and parents speak to babies in singsong musical baby talk. She thinks music is a special communication channel between parent child.[27]

Last but not least, my mother has told me that the reason she knew I could be worked with was that she realized I was humming Bach along with her while she was playing it on the piano. I was two years old and not talking, and I was doing things like ripping the wallpaper off the wall and eating it. I hadn't been diagnosed, but my mom knew something was drastically wrong, because I wasn't developing like the little girl next door who was my same age. But I could hum Bach.

All of these things make me believe there's a connection between music and language.

Scientifically speaking, I think we have some indirect support for this idea. DNA research on African tribes who speak *click languages,* languages in which the meaning comes from a change in tone, shows that *tonal languages* were probably the first language early humans spoke. Mandarin Chinese is also a tonal language. Tonal language isn't considered to be the same thing as music, but researchers who are studying nonnative speakers' ability to hear tone changes in Mandarin Chinese have found that music students out-perform nonmusic students.[28]

We also have good evidence that music developed in animals long before humans evolved. That evidence comes from a study of animal music by a pianist named Patricia Gray of the National Music Arts program and five biological scientists that was published in the prestigious journal *Science.* The authors write, "The fact that whale and human music have so much in common even though our evolutionary paths have not intersected for 60 million years, suggests that music may predate humans—that rather than being the inventors of music, we are latecomers to the musical scene."[29]

Animals are the originators of music and the true instructors. Humans probably learned music from animals, most likely from birds. More evidence that humans copied music from birds, rather than reinventing it for themselves: only 11 percent of all primate species sing songs.

Mozart was definitely influenced by birdsong. He owned a pet starling, and in his notebooks he recorded a passage from the Piano Concerto in G Major as he had written it, and as his pet starling had revised it. The bird had changed the sharps to flats. Mozart wrote, "That was beautiful" next to the starling's version. When his starling died, Mozart sang hymns beside its grave and read a poem he had written for the bird. His next composition, "A Musical Joke," has a starling style.[30] If a musical genius like Mozart admired and learned from a bird, it seems extremely likely early humans learned from birds when they were inventing the first human music.

Animal music is another case where human researchers are reluctant to say animals can do the same thing humans can do—animals can create music. Even Patricia Gray uses the phrase "musical sounds," not "animal music." Still, everyone agrees that individual elements of animal music are the same as individual elements of human music. Humpback whale songs contain repeating refrains the same way human songs do, and some whale songs rhyme. Whales probably use rhymes for the same reason people do, which is that rhymes help you remember what comes next in your poem or song. At Cornell University, Linda Guinee and Katy Payne (Katy Payne is the person who discovered that elephants use infrasonic sound to communicate) have found that long, complicated whalesongs are much more likely to rhyme than the shorter, easier songs.

Birds compose songs that use the same variation in rhythms and pitch relationships as human musicians, and can also transpose their songs into a different musical key. Birds use accelerandos, crescendos, and diminuendos, as well as many of the same scales composers use all over the world.

Animals and humans also have similar musical tastes. Rats and starlings can distinguish between "good" chords that sound consonant and dissonant chords that sound "bad." Luis Baptista, curator and chairman of the Department of Ornithology and Mammalogy at the California Academy of Sciences until his death in 2002, has a tape of a white-breasted wood wren in Mexico singing the exact opening notes of Beethoven's Fifth. It's unlikely that bird ever heard a recording of Beethoven's symphony before he sang it himself. The music that sounds beautiful to us also sounds beautiful to birds, and the bird composed the same theme.

Researchers also agree that animal song is highly complex, which makes it a good candidate for being a true animal language. Most animal communication researchers think animal calls are too simple to be real language. But nobody thinks animal song is simple. It could have the complexity to serve as a true animal language. To give just one example, it's likely that birds invented the sonata. A sonata begins with an opening theme, then changes that theme over the body of the piece, and finally ends with a repetition of the opening theme. Ordinary song sparrows compose and sing sonatas. A music psychologist named Diana Deutsch at the University of California at San Diego divides the sounds humans make into three categories: music, speech, and *paralinguistic utterances* like laughter or groans. She thinks animal calls are like our paralinguistic utterances but says, "When we come to birdsong, with its elaborate hierarchical patterning, it seems that [human] music provides a better analogy." In other words, animal music *is music.*[31]

Researchers who study animal songs say that animals use their songs to defend territory and attract mates, but I think animals probably use *tone language* to do more than that. We know music is deeply linked with emotions, because it lights up the emotional centers in the middle of the brain and even deep down in the cerebellum, which is the oldest part of the brain. A brain scan study by Carol Krumhansl of Cornell University found that music with a fast tempo played in a major key turned on the same physiological changes that happen when a person feels happy (such as faster breathing), while music in a minor key and a slow tempo produced the physiological changes that happen when you feel sad (slower pulse, higher blood pressure, drop in temperature).[32]

Maybe animals use tone to convey complex emotions to one another.

GIVING ANIMALS THE BENEFIT OF THE DOUBT

The fact is, we don't know very much about animal communication and animal language. If the history of animal research is anything to go on, we probably don't even know what we *think* we know, since every time researchers think they've proved animals can't do something along comes an animal who can. In animal communication and language, as in every other field of animal research, animals are going to turn out to be more capable than we know.

On the subject of animal communication, the debate comes down to two camps: people who think human language and animal communication are two separate and distinct things, and people who think human language and animal communication *are on the same spectrum.* Researchers who believe animal language is on a spectrum with human language believe that animal language might turn out to be simpler than human language, the way a two-year-old's language is simpler than a grown-up's, but it's still language. The difference is quantitative, not qualitative.

I vote with the spectrum people. I also believe animal researchers should change their paradigm. We've seen so many animals do so many remarkable things that it's time to start from the assumption that animals probably *do* have language rather than that they *don't*. The questions you ask set limits on the answers you find, and I think we'll learn more if we give animals the benefit of the doubt.

I'm going to end with a story about Alex. Dr. Pepperberg stays out of the language wars. She never says Alex has language, and she says she never will. I think that's probably more because she wants to stay out of the crossfire than because she thinks the language Alex has learned isn't "real." I say this because currently she is trying to see if Alex can disprove Noam Chomsky's latest proposal for what makes human language unique.

Noam Chomsky, Marc Hauser, and W. Tecumseh Fitch published an article in *Science* in 2002 arguing that humans are the only animals to have a language that is *recursive*. Loosely defined, recursive means that humans use rules to combine individual sounds and words into an infinite number of different sentences with different meanings.[33]

But Dr. Pepperberg points out that both dolphins and parrots can *understand* recursive sentences. Dolphins can handle sentences like "Touch the surfboard that is gray and to the left" versus "Swim over the Frisbee that is black and to your right." Apparently Noam Chomsky and his colleagues think that doesn't count, because the dolphins aren't creating these sentences; they're just understanding them. How any scientist can assume he knows *for a fact* that a dolphin doesn't produce recursive sentences in real life is a mystery to me.

Not very long ago, Dr. Pepperberg began trying to teach Alex and another gray parrot, Griffin, to sound out *phonemes,* which are the sounds that letters and letter combinations represent. English has forty phonemes altogether. She and her colleagues wanted to see if the birds understood that words are made out of letters that could be recombined to make other words, so they started training the birds with magnetic refrigerator letters.

One day their corporate sponsors were visiting Dr. Pepperberg's lab, and she and her staff wanted to show off what Alex and Griffin could do. So they put a bunch of colored plastic refrigerator letters on a tray and started asking Alex questions.

"Alex, what sound is blue?"

Alex made the sound "Sssss." That was right; the blue letter was "S."

Dr. Pepperberg said, "Good birdie," and Alex said, "Want a nut," because he was supposed to get a nut whenever he gave the right answer.

But Dr. Pepperberg didn't want him sitting there eating a nut during the limited time she had with their sponsors, so she told Alex to wait, and then asked, "What sound is green?"

The green example was the letter combination of "SH" and Alex said, "Ssshh." He was right again.

Dr. Pepperberg said, "Good parrot," and Alex said, "Want a nut."

But Dr. Pepperberg said, "Alex, wait. What sound is orange?"

Alex got that one right, too, and he *still* didn't get his nut. They just kept going on and on, making him sound out letters for his audience. Alex was obviously getting more frustrated by the minute.

Finally Alex lost his patience.

Here's the way Dr. Pepperberg describes it: Alex "gets very slitty-eyed and he looks at me and states, 'Want a nut. Nnn, uh, tuh.'"

Alex had spelled "nut." Dr. Pepperberg and her team were spending hours and hours training him on plastic refrigerator letters to see if Alex could eventually be taught that words are made out of sounds, and he already knew how to spell. He was miles ahead of them.

Dr. Pepperberg says, "These kinds of things don't happen in the lab on a daily basis, but when they do, they make you realize there's a lot more going on inside these little walnut-sized brains than you might at first imagine." I would like to add that there is a lot more going on than humans *perceive*. Dr. Pepperberg and her team are probably the world's foremost authorities on parrot cognitive abilities, they've been working with Alex for twenty years, and yet they had no idea Alex had learned to spell.[34]

—

It's time to start thinking about animals as capable and communicative beings. It's also time to stop making assumptions. Animal
researchers take a lot for granted: "animals don't have language,"
"animals don't have psychological self-awareness"—you find blanket
assertions like this sprinkled throughout the research literature. But
the truth is, we don't know what animals can't do any better than
we know what they can do. It's hard to prove a negative, and prov-
ing negatives shouldn't be the focus.

If we're interested in animals, then we need to study animals for
their own sake, and on their own terms, to the extent that it's possible. What are they doing? What are they feeling? What are they
thinking? What are they saying?

Who are they?

And: what do we need to do to treat animals fairly, responsibly,
and with kindness?

Those are the real questions.

7. Animal Genius: Extreme Talents

I t's getting to be obvious even to skeptics that animals are smarter than we think.

The question is, how much smarter?

My answer is that there are some animals who, like some people, have a form of genius. These animals have talents that are so extraordinary they're way past anything any normal human being could do even with a lot of hard work and practice.

Who are these animals?

Birds, for one. The more I learn about birds, the more I'm beginning to think we have no idea what the limits to some bird species' intelligence are. Bird migration is probably the most extraordinary talent we know about right now. Birds have brains no bigger than a walnut, but they can learn and remember migratory routes thousands of miles in distance. The Arctic tern has the longest migratory route we know about: 18,000 miles, round-trip. Some of these birds travel from the North Pole to the South Pole and back again every year.

EXTREME MEMORY

What makes this a genius-level ability instead of just some miraculous ability that's built into the species, like having wings and being able to fly, is the fact that birds have to *learn* these routes. They aren't born knowing their species' migratory route; it isn't hard-wired. Moreover, they learn the routes with almost no effort at all.

Many migratory birds have genius-level learning abilities when it comes to migration.

There's a good movie about these birds called *Fly Away Home,* based on the story of Bill Lishman, the man who, along with his partner, Joseph Duff, taught a bunch of Canada geese to follow him in his ultralight airplane. They created the project because they wanted to try to save the whooping cranes, which are on the verge of extinction. Operation Migration, the charity Bill Lishman founded, says there are only 188 whooping cranes left in the world. They're all in one big flock, which makes them even more vulnerable to extinction.

Up until Bill Lishman came along people were trying to save the species by raising baby whooping cranes in captivity. But it wasn't working because when the babies were brought up without any migrating adults to teach them the routes, there was no way to reintroduce them to the wild. They didn't know how to migrate, so when winter came they would just stay put and die in the cold.

Bill Lishman had the idea of teaching the whooping cranes to migrate by leading them along a migration path in his ultralight plane, a small one-person airplane that can fly as slowly as 28 to 58 miles per hour. He started out working with Canada geese, because geese aren't in danger of going extinct. Any golfer on the East Coast can tell you there's no goose shortage. As a matter of fact the goose poop problem has gotten so out of hand that some Border collies are getting a brand-new job working goose patrol at golf courses. That's good, because Border collies need a job. They get antsy living a life of leisure.

Pretty quickly Mr. Lishman managed to show that you could teach geese to follow a human in an ultralight airplane, *and* you could teach them a four-hundred-mile one-way migration route flying it just once. No human being could memorize a four-hundred-mile route across unmarked open terrain after traveling it just one time. Bird migration is an extreme talent.

After he knew he could do it with geese, he switched to sandhill cranes, which are related to whooping cranes but aren't endangered. In 1997 he led seven sandhill cranes from southern Ontario down to

Virginia, a four-hundred-mile trip one way. The cranes spent the winter in Virginia and then, one day at the end of March, they went out for their daily foraging and didn't come back. Two days later Mr. Lishman got a call from a school principal up in Ontario who said he had six big birds in his schoolyard entertaining the students! Six of the seven birds had made it the whole four hundred miles back to Canada, after having flown the route only once in their lives, and in the opposite direction. They ended up thirty miles away from where they'd been fledged.

Lots of animals have extreme memory and learning abilities in one realm or another. Gray squirrels bury hundreds of nuts every winter, one nut in each burial spot, and they remember them all. They remember where they hid each nut, what kind of nut it was, and even when they hid it. They're not just marking the spots some way, or finding the nuts by smell, which is what a lot of people probably assume. I read a gardening column the other day where a woman wrote in asking whether there was any way to keep squirrels from digging up her garden. The columnist answered that squirrels forget where they've buried their nuts, so they dig everything up. That is *not* true. Squirrels remember exactly where they buried hundreds and hundreds of nuts. Dr. Pierre Lavenex at the University of California, Berkeley, a researcher who studies memory in gray squirrels, says, "They use information from the environment, such as the relative position of trees and buildings, and they triangulate, relying on the angles and distances between these distant landmarks and their caches."[1]

No human can do that. A normal human can't even remember where he put the car keys half the time, let alone where he buried five hundred individual nuts. How long would a person last if he had to eat buried nuts for food? He wouldn't get through the winter, that's for sure. "People can do this [i.e. triangulate landmarks to find the precise spot where they've buried something] for a few sites," Dr. Lavenex says, "maybe six or seven, but not for nearly as many as squirrels do."

Most animals have "superhuman" skills like this: *animals have animal genius.* Birds are navigation geniuses, dogs are smell geniuses, eagles are visual geniuses—it can be anything.

EXTREME PERCEPTION
AND ANIMAL INTELLIGENCE

Many animals also have extreme perception. Forensic dogs are three times as good as any X-ray machine at sniffing out contraband, drugs, or explosives, and their overall success rate on tests is 90 percent.

The fact that a dog can smell things a person can't doesn't make him a genius; it just makes him a dog. Humans can see things dogs can't, but that doesn't make us smarter.

But when you look at the jobs some dogs have invented for themselves using their advanced perceptual abilities, you're moving into the realm of true cognition, which is solving a problem under novel conditions. The seizure alert dogs are an example of an animal using advanced perceptual abilities to solve a problem no dog was born knowing how to solve. Seizure alert dogs are dogs who, their owners say, can *predict* a seizure before it starts. There's still controversy over whether you can train a dog to predict seizures, and so far people haven't had a lot of luck trying. But there are a number of dogs who have figured it out on their own. These dogs were trained as seizure-response dogs, meaning they can help a person once a seizure has begun. The dog might be trained to lie on top of the person so he doesn't hurt himself, or bring the person his medicine or the telephone. Those are all standard helpful behaviors any dog can be trained to perform.

But some of these dogs have gone from responding to seizures to perceiving signs of a seizure ahead of time. No one knows how they do this, because the signs are invisible to people. No human being can look at someone who's about to have a seizure and see (or hear, smell, or feel) what's coming. Yet one study found that 10 percent of owners said their seizure response dogs had turned into seizure alert dogs.

The *New York Times* published a terrific article about a woman named Connie Standley, in Florida, who has two huge Bouvier des Flandres dogs who predict her seizures about thirty minutes ahead of time.[2] When they sense Ms. Standley is heading into a seizure they'll do things like pull on her clothes, bark at her, or drag on her hand to get her to someplace safe so she won't get hurt when the seizure begins. Ms. Standley says they predict about 80 percent of her seizures. Ms. Standley's dogs apparently were trained as seizure alert dogs before they came to her, but there aren't many dogs in that category. Most of the seizure alert dogs were trained to respond to seizures, not predict seizures.

The seizure alert dogs remind me of Clever Hans. Hans was the world-famous German horse in the early 1900s whose owner, Wilhelm von Osten, thought he could count. Herr von Osten could ask the horse questions like, "What's seven and five?" and Hans would tap out the number 12 with his hoof. Hans could even tap out answers to questions like, "If the eighth day of the month comes on Tuesday, what is the date for the following Friday?" He could answer mathematical questions posed to him by complete strangers, too.

Eventually a psychologist named Oskar Pfungst managed to show that Hans wasn't really counting. Instead, Hans was observing subtle, unconscious cues the humans had no idea they were giving off. He'd start tapping his foot when he could see it was time to start tapping; then he'd stop tapping his foot when he saw it was time to stop tapping. His questioners were making tiny, unconscious movements only Hans could see. The movements were so tiny the humans making them couldn't even *feel* them.

Dr. Pfungst couldn't see the movements, either, and he was looking for them. He finally solved the case by putting Hans's questioners out of view and having them ask Hans questions they didn't know the answers to themselves. It turned out Hans could answer questions only when the person asking the question was in plain view and already knew the answer. If either condition was missing, his performance fell apart.

Psychologists often use the Clever Hans story to show that humans who believe animals are intelligent are deluding themselves. But that's not the obvious conclusion as far as I'm concerned. No one has ever been able to *train* a horse to do what Hans did. Hans trained himself. Is the ability to read a member of a different species as well as Hans was reading human beings really a sign that he was just a "dumb animal" who'd been classically conditioned to stamp his hoof? I think there's more to it than that.

What makes Hans similar to the seizure alert dogs is that both Hans and the dogs acquired their skills without human help. As I mentioned, to my knowledge, so far no one's figured out how to take a "raw" dog and teach it how to predict seizures. About the best a trainer can do is reward the dogs for helping when a person is having a seizure and then leave it up to the dog to start identifying signs that predict the onset of a seizure on his own. That approach hasn't been hugely successful, but some dogs do it. I think those dogs are showing superior intelligence the same way a human who can do something few other people can do shows superior intelligence.

What makes the actions of the seizure alert dogs, and probably of Hans, too, a sign of high intelligence—or high talent—is the fact that they didn't have to do what they did. It's one thing for a dog to start recognizing the signs that a seizure is coming; you might chalk that up to unique aspects of canine hearing, smell, or vision, like the fact that a dog can hear a dog whistle while a human can't. But it's another thing for a dog to start to recognize the signs of an impending seizure and *then decide to do something about it*. That's what intelligence is in humans; intelligence is people using their built-in perceptual and cognitive skills to achieve useful and sometimes remarkable goals.

INVISIBLE TO THE NAKED EYE

By now you're probably thinking, if animals are so smart, why hasn't anyone noticed?

First of all, we have no idea what most animals are doing in the wild. Even when people like Jane Goodall have been able to spend years doing close observation of a group of animals in their native habitat, we still don't learn what the *animals* think they're doing, or what they're communicating to one another about what they're doing. That's why it's always a surprise when a crow like Betty spontaneously bends a wire to make a food hook, or a gray parrot like Alex suddenly spells the word "nut." Just the other day I met a lady at a conference who told me about another super-smart bird living in a Florida hotel. This bird is a macaw who invented a new word—

crackey—to signify either cookie or cracker. Those are the two foods his owner gives him as treats, so apparently the macaw decided that *cookie-cracker* is a food category unto itself, requiring its own word, which he created by putting "cookie" and "cracker" together. He's right about cookies and crackers; they *are* a separate category. Cookies and crackers are both treats, not "real" food. I'm guessing that's what the bird means when he asks for a crackey; he's probably asking for junk food.

Another gray parrot, N'Kisi, owned by Aimee Morgana in New York City, has a vocabulary of over five hundred English words. She uses the present, past, and future tenses and once used the word "flied" to mean "flew." She called the aromatherapy oils Aimee uses "pretty smell medicine."

The point is, we don't know what animals can and can't do. The fact that we're constantly being dumbfounded by brand-new abilities no one had a clue animals possessed ought to be a lesson to us about how much we don't know.

IF ANIMALS ARE SO SMART, WHY AREN'T THEY IN CHARGE?

I think the reason researchers don't take this lesson more to heart is that most people just naturally assume, without stopping to think about it, that if animals were as smart as humans or smarter, they'd have more to show for it. Where are all the animal inventions? That's the big question.

This is the if-animals-were-smart-they-wouldn't-still-be-pooping-in-the-woods theory of animal cognition. If animals were *really* smart, they would have invented flush toilets!

What the indoor plumbing theory of animal IQ forgets is the fact that plenty of indigenous peoples never invented indoor plumbing, either, and they're no less intelligent than anyone else. Our thinking about animals is a lot like the Europeans' thinking about primitive cultures in the nineteenth century when European explorers first began to have a lot of contact with the people of Africa. That was a time when botanists and zoologists were creating classifications for every plant and animal on earth, so naturally Europeans created classifications for humans, too. They thought the Europeans were the most intelligent, the Asians were next most intelligent, and the Africans were on the bottom.

The Europeans were wrong about that, probably for some of the same reasons people will turn out to be wrong about animals, too. One big mistake the Europeans made was to equate *IQ* with *cultural evolution*. *Cumulative cultural evolution* means that each generation can build on the knowledge of the generation before it rather than having to start all over again from scratch. For a culture to evolve, you have to have *cultural ratcheting*, which means that a group of people or animals has to have a way to hold on to the things the previous generations have learned so the next generation can add on new things.[3] Cultural ratcheting means a culture can maintain and pass along an expanding body of knowledge that no one generation would be able to invent for itself.

Researchers don't know how and why one culture evolves faster than another, but they do know it's not because of IQ. You probably have to have things like direct, one-on-one teaching along with *very* widespread paying attention and learning so you don't keep losing knowledge as fast as you gain it.

All human cultures, including indigenous peoples, have *cumulative cultural evolution* to some degree. But so far researchers think only birds and *maybe* chimpanzees also have it. However, there is so much of animal life we just can't perceive at this point, that the time hasn't come to conclude that animals do or do not have cultural evolution. Take dolphins, for instance. Dolphins talk back and forth to each other for hours on end. It's completely possible dolphins could have a rich "mental" culture they've developed over many generations that's invisible to us. How would we know one way or the other?

I thought about dolphins when I read *A Man Without Words*. In deaf culture people sign the same information to each other over and over again to make sure every person understands it and has the same information. The author, Susan Schaller, talks about a picnic she attended where "even though everyone saw my name and where I was from in my [signed] introduction, the spelling of my English name, my namesign, and California's namesign passed from person to person until everyone was completely satisfied that they had all seen the exact same information."

I wonder whether dolphins are doing something like that, passing precious cultural information from dolphin to dolphin over and over again to make sure none of it gets lost. Dolphins don't have books or hands, so they can't record the things they know in writing *or* in objects they've built. I say this because early humans didn't have written language, either, but they made simple tools, clothing, and shelters that could probably serve both as objects *and* as the instructions on how to make the object. (When an object is really simple, you can tell a lot about how to make it just by looking at it.)

But if you have only oral communication, and you've built a complex culture, then passing your culture along would be like playing the game Telephone. You'd be constantly in danger of having distortions come into the transmission process, ruining the knowledge you're trying to pass along. The only way to keep this from happening would be to develop a strict habit of repeating each piece of knowledge over and over again, back and forth, to make sure the person or dolphin you're transmitting to has received an exact copy of your message, not an approximation.

Smart, but Different

I think animals are smarter than we know. I also think a lot of animals probably have a different *kind* of intelligence than g, the general fluid intelligence normal people have.

In the last chapter I said that animals are cognitive specialists. They're smart in some things, not smart in others. People are generalists, meaning that a person who's smart in one area will be smart in others, too. That's what IQ tests show.

Autistic people are smart the way animals are smart. We're specialists. Autistic people can have IQ scores all over the map. Donna Williams, an autistic woman from Australia who wrote a memoir called *Nobody Nowhere,* has written that her own scores on the different subscales range all the way from mentally retarded to genius. I believe it.[4]

After many years observing animals and living with autism, I have come to the conclusion that animals with extreme talents are similar to autistic savants.

If you've never met an autistic savant, you might want to watch the movie *Rain Man,* which is about an autistic savant, Raymond, and his brother. Raymond couldn't fix himself a piece of toast without setting the kitchen on fire, but he could count cards in a game of blackjack and win thousands of dollars. That kind of disparity is typical with autistic savants. When you get outside their specialty they're almost never as smart or capable as normal people. That's why they used to be called *idiot savants.* Just like animals with extreme talent, autistic savants can *naturally* do things no normal human being can even be *taught* to do, no matter how hard he tries to learn or how much time he spends practicing. Yet they usually have IQs in the mentally retarded range.

LUMPERS AND SPLITTERS: WHAT MAKES ANIMALS AND AUTISTIC PEOPLE DIFFERENT

Charles Darwin first used the terms *lumpers* and *splitters* to describe the two different kinds of taxonomists. Lumper taxonomists grouped lots of animals or plants into big, broad categories based on major characteristics; splitters divided them up into lots of smaller categories based on minor variations. Lumpers generalize; splitters "particularize."

This is a core difference between animals and autistic people on the one hand, and normal people on the other. Animals and autistic people are splitters. They see the differences between things more than the similarities. In practice this means animals don't generalize very well. (Normal people often over-generalize, of course.) That's why you have to be so careful when you're socializing an animal to socialize him to many different animals and people.

You have to do the same thing with training. Service dogs who are being trained to lead a blind person across the street don't generalize from one intersection to another, so you can't just train them on a couple of intersections and expect them to apply what they've learned to a brand-new intersection. You have to train them on dozens of different kinds of intersections: corners where there's a light hanging in the middle of the intersection and crosswalk lines painted on the pavement, corners where there's a light hanging in the middle of the intersection and no crosswalk lines, corners where the traffic lights are on poles, and so on.

This is why dog trainers always make people train their own dogs. You can't send a puppy away to obedience school, because he'll only learn to obey the trainer, not you. Dogs also need some training from every member of the household, because if only one person trains the dog, that's the only person the dog is going to obey.

And you have to be careful not to fall into *pattern training*. Pattern training happens when you always train the dog in the same place at the same time using the same commands in the same order. If you pattern-train a dog, he'll learn the commands beautifully, but he won't be able to perform them anyplace other than the spot you trained him in, or in any sequence other than the one you used during training. He's learned the pattern, and he can't generalize the individual commands to other times, settings, or people.

People who teach autistic children deal with exactly the same challenge. A behaviorist told me a story about an autistic boy he'd been teaching how to butter toast. The behaviorist and the parents had been working really hard with the boy, and finally he got it. He could butter toast. Everyone was thrilled, but the joy didn't last too long, because when somebody gave the boy some peanut butter to spread on his toast, he didn't have a clue! His brand-new bread-buttering skill was specific to butter, and it didn't generalize to peanut butter. They had to start all over again and teach him how to spread *peanut* butter on toast. This happens all the time with autistic people, and with animals, too.

It happens so much, and it's so extreme, that it's not right just to call animals splitters; animals are *super-splitters*. That's what being hyper-specific is all about.

It's not that animals and autistic people don't generalize at all. Obviously they do. The black hat horse generalized his original traumatic experience to other people wearing other black hats, and the little boy who could butter toast had generalized that skill to other sticks of butter and other pieces of bread. With training, a service dog learns to generalize what he knows about other intersections to new intersections he's never seen before.

What's different is that the generalizations animals and autistic people make are almost always narrower and more specific than the generalizations nonautistic people make. *Human with black hat* or *spread butter on bread*: those are pretty narrow categories.

THE HIDDEN FIGURES TALENT

To any normal person, being hyper-specific sounds like a serious mental handicap, and in a lot of ways it is. Hyper-specificity is probably the main reason animals seem less smart than people. How intelligent could a horse be if he thinks the really scary thing in life isn't a nasty handler but the nasty handler's hat?

Probably not too intelligent when it comes to school smarts. But being smart in school isn't everything, and high general intelligence comes at the price of high *hyper-specific intelligence*. You can't have both.[5]

That means normal human beings can't have extreme perception the way normal animals can, because hyper-specificity and extreme perception go together. I don't know whether one causes the other, or whether hyper-specificity and extreme perception are just different aspects of the same difference in the brain. What I do know is that Clever Hans couldn't do what people do, and people can't do what Hans did. Hans had a special talent humans don't have.

Until we know more about it, I'm calling this ability the *hidden figure talent*, based on some research findings in autism. In 1983 Amitta Shah and her colleague Uta Frith tested twenty autistic children, twenty normal children, and twenty children with learning disabilities—all of them the same mental age—on the Embedded Figure Task. In the test, first you show the child a shape, like a triangle, and then you ask him to find the same shape inside a picture of an object like a baby carriage.

The autistic children did much better at finding the hidden figure than any of the other children. They almost always saw the figure instantly, and they scored 21 out of 25 correct answers on average, compared to an average of only 15 correct answers for both the learning disabled and the normal kids. That's a huge difference. It's so huge you could probably say normal people are disabled compared to autistic people when it comes to finding hidden figures. The autistic children were so good they almost outscored the experimenters! These were developmentally disabled kids scoring the same as normal adults.[6]

I believe it, because a few years back I happened to come across a hidden figure test in *Wired Magazine,* and the hidden figures jumped out at me. For me, they weren't really hidden.

To my knowledge no one's ever tested animals on hidden figure tests, but I bet they'd do well. Probably the easiest way to do a hidden figure test with an animal would be to run a simple recognition task. Teach the animal to touch or peck a certain shape, then show him a picture with the shape embedded inside and see whether the animal can still find it.

Most people don't realize how valuable the hidden figure talent is in the right situation. In Maryland there's an employment agency for autistic adults that places its clients in jobs like quality assurance. They have one group of autistic men working in a factory inspecting logo T-shirts coming off the line for flaws in the silk-screening. Nonautistic people have a hard time seeing tiny differences between one silk-screened logo and another, but those autistic employees can pick up practically microscopic flaws in a glance. It's the hidden figure test all over again. To them the flaws in the silk-screening aren't hidden.

The agency's clients also outperform normal people in bindery work. When you're assembling corporate reports you have to be able to tell the front cover from the back cover quickly and accurately. To regular people the fronts and backs look alike, but autistic employees can always tell the front from the back, and they do it in a flash. Extreme perception lets them see all the tiny differences normal people can't see. The agency even has one autistic woman working quality assurance on submarine parts.

I thought about those employees not too long after 9/11 when news reports started coming out about how hard it is for people who work as luggage inspectors to spot weapons on their video screens due to *clutter.* If you're a normal human being and your job is to sit in one place all day long staring at a video screen, pretty soon you'll have trouble separating out the form of a weapon from all the other junk that's packed in people's bags. The screen is too cluttered, and everything blurs together. But that might not be a problem for autistic people, and I think airports ought to try out some autistic people in that job.

I think we're letting a huge amount of talent go to waste, both in people who aren't "normal" and in animals who are. That's probably because we don't really understand what animals could do if we gave them a chance. We're just leaving it up to animals like the seizure alert dogs to invent their own jobs.

AUTISTIC SAVANTS

I mentioned at the beginning of this book that I think animal genius is probably the same thing as autistic savantry. I've felt this way for years, just from being around animals and observing them, and I mentioned it in *Thinking in Pictures*. But I didn't know why autistic genius and animal genius looked so similar to me, or whether autistic genius and animal genius might come from the same difference in the brain.

It's not that autistic savants and *animal savants* do the same things. Animal savants show brilliance when they learn complicated migratory routes after just one flight or discover how to perceive seizures before they happen. Autistic savants do lightning-fast calendar or prime number calculations inside their heads, or become artistic savants who can make almost perfect line drawings of buildings and landscapes from memory, often starting from a very young age—*and using perfect perspective*. That's especially amazing, because even great artists have to be taught how to draw using perspective. A four-year-old autistic savant just naturally knows how to do it.

Even though autistic savantry and animal savantry seem so different on the surface, the one thing that did jump out was that a lot of these talents involve amazing feats of rote memory. Autistic people are known for their ability to memorize whole train schedules, the capitals of every country in the world, and so on. Autistic savants are the only people who seem like they could give a Clark's nutcracker a run for its money when it comes to remembering where they hid thirty thousand pine seeds. But beyond that, I didn't know why animal genius felt so familiar to me.

Then in 1999 Dr. Allan Snyder, a psychologist at the Centre for the Mind at Australian National University, published a paper that laid out a *unified theory* of all the different savant talents. If his theory is right, it probably explains animal genius, too.[7] Dr. Snyder and his co-author, Dr. D. John Mitchell, say that *all* the different autistic savant abilities come from the fact that autistic people don't process what they see and hear into unified wholes, or *concepts,* rapidly the way normal people do.

A normal person looks at a building and his brain turns all the hundreds and thousands of building pieces coming in through his sensory channels into one unified thing, a building. The brain does this automatically; a normal person can't *not* do it. That's why a common drawing lesson art teachers use is to have art students turn a picture upside down and copy it that way, or else draw the *negative space* surrounding an object instead of the object itself. Turning the object upside down or drawing the negative space tricks your brain into letting the image stay in separate pieces more easily,[8] so you can draw the object instead of your *unified concept* of the object. People are always amazed at how good their upside-down drawings are.

Autistic people are stuck in the *pieces* stage of perception to a greater or lesser degree, depending on the person. Donna Williams, the autistic woman who wrote the book *Nobody Nowhere,* says she can't really see a whole object all at once. She sees a kind of slide show of the object. If she's looking at a tree, first she might see a branch on that tree, then the screen changes and she sees a bird sitting on the branch, then the screen changes again and she sees some leaves, and so on. Some autistic people have this problem a lot worse than others, and I think it's possible some autistic people have such fragmented sensory systems that they may be almost blind or deaf. I wonder whether some autistic people are so deprived of coherent sensory input that they are like autistic Helen Kellers.

Snyder and Mitchell say that the reason autistic people see the pieces of things is that they have *privileged access* to *lower levels of raw information.* A normal person doesn't become conscious of what he's looking at until after his brain has composed the sensory bits and pieces into wholes. An autistic savant is conscious of the bits and pieces.

That's why autistic savants can make perspective drawings without being taught how. They're drawing what they see, which is all the little changes in size and texture that tell you one object is closer up and another object is farther away. Normal people can't see all those little changes without a lot of training and effort, because their brains process them unconsciously. So normal people are drawing what *they* "see," which is the finished object, after their brains have put it all together. Normal people don't draw a dog, they draw a *concept* of a dog. Autistic people draw the dog.

It's ironic that we always say autistic children are in their own little worlds, because if Dr. Snyder is right it's normal people who are living inside their heads. Autistic people are experiencing the actual world much more directly and accurately than normal people, with all their inattentional blindness and their change blindness and their every-other-kind-of-blindness. (Dr. Snyder hasn't talked about inattentional or change blindness that I know of, but the research on those concepts supports his work.)

Math savants use this same brain difference to do calendar calculations and prime number identification. An autistic savant who can tell you on what day you were born is seeing time as a sequence of seven different days repeating over and over again going back to the beginning of time. They quickly scan back over the pattern until they come to your day.

Normal people don't experience time that way. To a normal person a month or a year or a decade is one unified time span, not a collection of separate and distinct days. It's a blur. (Dr. Snyder's theory is a little more complicated than I've been making it sound. He thinks the brain has a processor that divides all incoming data—time, space, objects, and so forth—into equal parts. That's why an autistic savant can tell whether a number is prime or not, because a prime number *can't* be divided.)

Calendar calculation is the hidden figure talent all over again. I believe most or even all of the savant talents autistic people have are variations on the hidden figure ability.

I also believe that most or even all of the savant talents animals have are variations on the hidden figure ability, and in just the past couple of years Dr. Snyder and Dr. Bruce Miller, a physician at the University of

California at San Francisco, have supplied some hard evidence that I may be right. Dr. Miller works with patients who have a disorder called *frontotemporal dementia* in which the front part of the brain progressively loses its functions. In frontotemporal dementia the frontal lobes and the temporal lobes, which are at the side of your head, are affected.[9] Neither of these areas is working well in autistic people either, and as I've been saying throughout this book, the biggest area of difference between the animal brain and the human brain is that an animal's frontal lobes are smaller and less well developed than a human's. Serious frontal lobe damage is worse than being autistic. If your frontal lobes are badly damaged you can have symptoms of practically all the psychiatric disorders—autism, ADHD, obsessive-compulsive disorder, severe mood disorders, you name it.

You're probably going to have at least *some* autistic symptoms. We know that Dr. Miller's patients do, because some of them start to develop savant talents. A few of these people have become artists in their fifties and sixties, even winning awards in art shows. Others have developed musical abilities; one patient invented a chemical detector and got a patent for it. When he made his invention he could name only one out of fifteen objects on a standardized word test. A patient who had lost *all* his language ability designed sprinklers! These patients had *sudden-onset* talents.

I suspect what's happening with these people is that all of a sudden they're able to have the same kind of hyper-specific perception that underlies an autistic savant's ability to do a calendar calculation or make a perspective drawing without being taught.

Dr. Snyder has now begun to test the proposition that savant talents come from conscious access to the raw data of the brain. When he uses magnetic stimulation to interfere with frontal lobe functioning in his subjects, they start to make much more detailed drawings than they could just moments before.[10] They also get better at proofreading. Before he turns on the magnetic stimulation, Dr. Snyder has his subjects read this poem out loud:

> *A bird in the hand*
> *is worth two in the*
> *the bush*

Almost all people look at the poem and say, "A bird in the hand is worth two in the bush."

About five minutes after he turns on the magnetic stimulation some of his subjects suddenly read, "A bird in the hand is worth two in the *the* bush." The duplicate "the" pops out at them as their left frontal-temporal lobes go down, and they start turning into hidden figure specialists, perceiving detail they didn't perceive before. One of them even told Dr. Snyder that he felt more "alert" and "conscious of detail." He was so intensely aware of the details around him that he said he wished they had asked him to write an essay, something he normally didn't like to do.

THE DEVIL IS IN THE DETAILS

I don't know whether extreme talents in animals work the same way Dr. Snyder thinks they work in people with autism, but we have a lot of evidence that animals at least *see* the world in sharper detail than regular people do. I've already talked about how important visual detail is to animals, but we also have some fascinating research on ant navigation that goes along with Dr. Snyder's experiments.

When ants walk through an obstacle course they use landmarks to remember their route the same way people do. If they pass a gray pebble going one way, they'll look for that same gray pebble coming back.

But there's one big difference. When an ant reaches a landmark, he does something normal people don't do. He passes the landmark, stops, turns around, and *looks at the landmark from the same spot where he saw it on the trip out.*

He has to do that, because to an ant a gray pebble probably looks different coming and going. He has to see the pebble from the same vantage point where he saw it first to make sure it's still the same gray pebble he saw before. This says to me ants probably don't automatically combine separate pieces of sensory data into wholes in the same way or to the same degree normal humans do.

For a nonautistic person, a landmark looks the same coming or going. When a normal person sees a big red barn on the way to someone's house, he automatically sees the same big red barn on the way back. It looks the same to him, even though he's seeing it from a different side.

That's because a normal person's nervous system gets rid of a lot of detail and then fills in the blanks with whatever he *expects* to see. If he were consciously seeing what's really in front of his eyes, he'd see a slightly different red barn coming and going, because the south side of a barn doesn't look exactly like the north side of a barn, and the east side doesn't look exactly like the west side. Even if the builder designed all four sides to be identical, in nature there's always a difference in light and shadow.

I do the same thing ants do, which is one more thing that makes me think hyper-specificity is a key link between animals and autistic people. When I drive someplace I've never been before I look for landmarks along the road the same way everyone else does. But then when I'm driving back, the landmarks I've picked out all look different to me. I have to drive past each landmark until I reach the spot where I was when I first saw it; then I turn around and look at it from the original angle to make sure it's the same thing I saw on my way out. For animals and for people with autism, different sides of the same object actually *look different*.

THINKING ABOUT WHAT ANIMALS CAN DO, NOT WHAT THEY CAN'T

I hope we'll start to think more about what animals *can* do, and less about what they can't. It's important, because we've gotten too far away from the animals who should be our partners in life, not just pets or objects of study.

You always hear that humans domesticated animals, that we turned wolves into dogs. But new research shows that wolves probably domesticated people, too. Humans *co-evolved* with wolves; we changed them and they changed us.

The story of how researchers have begun to piece this together is an example of *converging lines of evidence*, which is what happens when findings from different fields start to fit together and all point in the same direction. For a long time, the best evidence researchers had about when and how wolves turned into dogs came from archaeological discoveries of dog remains that had been carefully buried underneath humans' huts. Some archaeologists found dogs and people buried together in the same grave.

Those first buried dogs date back about 14,000 years. Humans had not yet invented farming at that time, but they had the same bodies and brains we do. So it made sense to conclude that primitive humans evolved into modern humans first, then began to associate with wild wolves who subsequently evolved into the domestic dog, in order to serve as working dogs and pets.

But a study by Robert K. Wayne and his colleagues at UCLA of DNA variability in dogs found that dogs had to have diverged from wolves as a separate population 135,000 years ago.[11] The reason the fossil record doesn't show any dogs with humans before 14,000 years ago is probably that before then people were partnered with wolves, or with wolves that were evolving into dogs. Sure enough, fossil records do show lots of wolf bones close to human bones before 100,000 years ago.

If Dr. Wayne is right, wolves and people were together at the point when *homo sapiens* had just barely evolved from *homo erectus*. When wolves and humans first joined together people only had a few rough tools to their name, and they lived in very small nomadic bands that probably weren't any more socially complicated than a band of chimpanzees. Some researchers think these early humans may not even have had language.

This means that when wolves and people first started keeping company they were on a lot more equal footing than dogs and people are today. Basically, two different species with complementary skills teamed up together, something that had never happened before and has really never happened since.

Going over all the evidence, a group of Australian anthropologists believes that during all those years when early humans were associating with wolves *they learned to act and think like wolves.*[12] Wolves hunted in groups; humans didn't. Wolves had complex social structures; humans didn't. Wolves had loyal same-sex and nonkin friendships; humans probably didn't, judging by the lack of same-sex and nonkin friendships in every other primate species today. (The main relationship for chimpanzees is parent-child.) Wolves were highly territorial; humans probably weren't—again, judging by how nonterritorial all other primates are today.

By the time these early people became truly modern, they had learned to do all these wolfie things. When you think about how different we are from other primates, you see how doglike we are. A lot of the things we do that the other primates don't are dog things. The Australian group thinks it was the dogs who showed us how.

They take their line of reasoning even further. Wolves, and then dogs, gave early humans a huge survival advantage, they say, by serving as lookouts and guards, and by making it possible for humans to hunt big game in groups instead of hunting small prey as individuals. Given everything wolves did for early man, dogs were probably a big reason why early man survived and Neanderthals didn't. Neanderthals didn't have dogs.

But dogs didn't just help people stay alive long enough to reproduce. Dogs probably also made it possible for humans to pull ahead of all their primate cousins. Paul Tacon, principal research scientist at the Australian Museum, says that the development of human friendship "was a tremendous survival advantage because that speeds up the exchange of ideas between groups of people." All cultural evolution is based on cooperation, and humans learned from dogs how to cooperate with people they aren't related to.[13]

Maybe the most amazing new finding is that wolves didn't just teach us a lot of useful new behaviors. Wolves probably also changed the structure of our brains. Fossil records show that whenever a species becomes domesticated its brain gets smaller. The horse's brain shrank by 16 percent; the pig's brain shrank as much as 34 percent; and the dog's brain shrank 10 to 30 percent. This probably happened because once humans started to take care of these animals, they no longer needed various brain functions in order to survive. I don't know what functions they lost, but I do know all domestic animals have reduced fear and anxiety compared to wild animals.

Now archaeologists have discovered that 10,000 years ago, just at the point when humans began to give their dogs formal burials, the human brain began to shrink, too. It shrank by 10 percent, just like the dog's brain. And what's interesting is what *part* of the human brain shrank. In all of the domestic animals the *forebrain,* which holds the frontal lobes, and the *corpus callosum,* which is the connecting tissue between the two sides of the brain, shrank. But in humans it was the *midbrain,* which handles emotions and sensory data, and the *olfactory bulbs,* which handle smell, that got smaller while the corpus callosum and the forebrain stayed pretty much the same. Dog brains and human brains specialized: humans took over the planning and organizing tasks, and dogs took over the sensory tasks. Dogs and people coevolved and became even better partners, allies, and friends.

"Dogs Make Us Human"

The Aborigines have a saying: "Dogs make us human." Now we know that's probably literally true. People wouldn't have become who we are today if we hadn't co-evolved with dogs.

I think it's also true, though in a different way, that *all* animals make us human. That's why I hope we'll start to think more respectfully about animal intelligence and talent. That would be good for people, because there are a lot of things we can't do that animals can. We could use their help.

But it would be good for animals, too. Dogs first started living with people because people needed dogs and dogs needed people. Now dogs still need people, but people have forgotten how much they need dogs for anything besides love and companionship. That's probably okay for a dog who's been bred to be a companion animal, but a lot of the bigger breeds and practically all of the mixed breeds were built for work. Having a job to do is part of their nature; it's who they are. The sad thing is, now that hardly anyone makes his living herding sheep, most dogs are out of a job.

It doesn't have to be that way. I read a little story on the Web site for the American Veterinary Medical Association that shows the incredible things animals are capable of doing, and would do if we gave them the chance. It was about a dog named Max who had trained himself to monitor his mistress's blood sugar levels even while she was asleep. No one knows how Max was doing this, but my guess is people must smell slightly different when their blood sugar is low, and Max had figured that out.[14] The lady who owned him was a severe diabetic, and if her blood sugar levels got low during the night Max would wake up her husband and bug him until he got up and took care of her.

You have to think about that story for only five seconds to realize how much dogs have to offer. Dogs and a lot of other animals.

—

People always wonder how I can work in the meatpacking industry when I love animals so much. I've thought about this a lot.

After I developed my center-track restraining system, I remember looking out over the cattle yard at the hundreds and hundreds of animals milling around in their corrals. I was upset that I had just designed a really efficient slaughter plant. Cows are the animals I love best.

Looking at those animals I realized that none of them would even exist if human beings hadn't bred them into being. And ever since that moment I've believed that we brought these animals here, so we're responsible for them. We owe them a decent life and a decent death, and their lives should be as low-stress as possible. That's my job.

Now I'm writing this book because I wish animals could have more than just a low-stress life and a quick, painless death. I wish animals could have a *good* life, too, with something useful to do. I think we owe them that.

I don't know if people will ever be able to talk to animals the way Doctor Doolittle could, or whether animals will be able to talk back. Maybe science will have something to say about that.

But I do know people can learn to "talk" to animals, and to hear what animals have to say, better than they do now. I also know that a lot of times people who can talk to animals are happier than people who can't. People were animals, too, once, and when we turned into human beings we gave something up. Being close to animals brings some of it back.

Behavior and Training Troubleshooting Guide

Training, solving behavior problems, and understanding why animals do what they do will be easier if you know the motivations for different behaviors.

Animal behavior is a complex mixture of learned behaviors, biologically based emotion, and hardwired instinctual behavior.

Examples of hardwired behaviors are bird mating dances and a dog chasing something that moves rapidly. Ethologists call these behaviors, which are always the same and never vary from one member of a species to another, fixed action patterns. Fixed action patterns are turned on by sign stimuli. The sign stimulus for prey chasing is rapid movement, while a bird's mating dance is triggered by the sight of a potential mate as well as a surge of hormones.

The fixed action pattern is hardwired but the particular sign stimulus that turns it on is determined by learning and emotion. A basic principle of animal behavior is that WHO you have sex with, WHAT you eat, WHERE you eat, WHO you fight with, and WHO you socialize with are *learned*. In dogs the killing bite is instinctual, but the animal learns what to kill and what not to kill. Chasing things that move rapidly is instinctual, but a dog learns that he *can* chase a ball but he *cannot* chase children.

Brain research now shows that the way the brain processes various core motivations, or emotions, is different. Example: fear and rage are neurologically very different. Being scared and being angry are two different feelings. Both humans and mammals have similar systems in the brain for processing basic emotions.

Another important principle in shaping animal behavior is the fact that animals are individuals. One dog may have high social motivation and respond well to praise alone. Another dog may be more motivated by food rewards. The degree of fearfulness varies greatly between different breeds of animals, but the range of fearfulness within the same breed may also vary greatly. On average, Arab horses and Border collies have higher fearfulness than quarter horses and Rottweilers, but there will be some low-fear Arabs who will have the lower fear levels of a quarter horse.

A high-fear horse or dog is more likely to be traumatized by abuse than a low-fear animal. An Arab horse who has been beaten may become so fearful that he becomes dangerous to ride, but a horse with a less fearful temperament may habituate to some rough treatment. Dogs, cats, and other animals with the high-fear trait often quiver and tremble, and are more likely to panic and startle when they are exposed to a sudden novel stimulus such as an umbrella opening or a piece of metal falling on the floor. A horse who is getting scared will hold his head up high, switch his tail, and sweat. A high-fear dog who has been abused will often cower and crouch down when a person approaches. He may also bite, especially if he feels cornered.

Anyone training an animal should rely on positive emotions and motivations, such as praise, stroking, or food rewards, not negative. Animals learn new skills more easily with positive reinforcement, and learning new behaviors should always be a good experience for the animal.

Below is a list of the basic *behaviors* and *behavior motivators* in animals:

1. Fear
2. Rage and anger
3. Predatory chasing
4. Sociality
5. Pain
6. Novelty seeking and novelty avoidance
7. Hunger
8. Sex

BEHAVIORS AND BEHAVIOR MOTIVATORS
Fear-Motivated Behaviors
Examples

- An animal struggles and vocalizes during veterinary procedures.
- A stressed-out dog bites at a noisy party.
- A horse rears when he sees a person wearing a black hat because he was once abused by someone wearing a black hat.
- A horse is calm when ridden at home but goes berserk at a show when he sees balloons for the first time.
- An abused dog cowers and bites when a person raises his hand.
- A horse kicks (spooks) at blowing paper.
- A dog runs under the couch during a thunderstorm.
- A cat goes berserk at a veterinary clinic when he sees a dog for the first time. This is an example of extreme fear and panic.
- A monkey at a zoo runs and hides when he hears the voice of the person who immobilized him with a tranquilizer dart.

- A dog growls at men because he was abused by a man.
- A horse goes berserk when ridden with a type of bit used to abuse him. Changing the bit to one that feels totally different will sometimes prevent this fear-motivated behavior.
- A horse bucks when he changes gaits. This often occurs when training has been done too rapidly. During too rapid training, a horse can be frightened by the new sensations of the saddle when he changes from a walk to a different gait. Trying a different pad and saddle that will feel completely different to the horse may help. Once the different pad and saddle are in place, allow the horse to *gradually* become accustomed to how they feel on his back at different gaits.
- A horse refuses to load onto a trailer because he hit his head the first time he entered one.
- A horse bites for no apparent reason. This often occurs in a horse who has been abused or subjected to harsh training methods.

Principles of Troubleshooting

- Never punish fear-motivated behavior because the animal will become increasingly frightened.
- Fear-motivated behaviors are more likely to occur in high-strung flighty animals. Flightiness and the tendency to startle easily are inherited, genetically based traits. On the average, horses as a species have more fear-motivated behavior than dogs. Some breeds are more flighty than others, but there can be great variation. In all species, animals with fine small bones and slender bodies are usually more fearful than heavier-boned animals with heavier bodies. Horses and cattle with a spiral hair whorl above the eyes are more likely to be flighty compared to ones with a spiral hair whorl located below the eyes.
- Fear-motivated behaviors often occurs in abused animals.
- Gentle, positive training methods can often prevent fear-motivated behavior. This is especially important with high-fear flighty animals.
- Frightened animals are easier to handle if they are given twenty to thirty minutes to calm down.
- Animals with high-fear genetics such as Arab horses, Border collies, and many of the small dog breeds are more likely to be traumatized and damaged by harsh treatment.
- Use calming methods to soothe the animal such as stroking and talking in a calm low voice. Stroke the animal, do not pat him. Some animals interpret pats as hits.
- A frightened animal will often calm down and relax when he hears the calm voice of a familiar trusted person. It is advisable to train animals to trust more than one person.
- Trainers must work to prevent fear memories from forming, especially

in nervous flighty animals. An animal's first experiences with a new person, place, or piece of equipment should be positive. Example: if a horse falls down the first time he is loaded into a trailer he may develop a fear of trailers.

- Fear memories are permanent. Since animals do not have language, fear memories are stored as pictures, sounds, touch sensations, or smells. An animal may become scared if he or she sees, hears, touches, or smells something that is associated with a painful or frightening experience.

- Although fear memories are permanent, desensitization programs can help. Example: a dog's fear of thunderstorms may be able to be desensitized by playing a sound effect recording of a thunderstorm with gradually increasing volume.

- Some fear-motivated behaviors may require a veterinarian's prescription for either antidepressant medications or anti-anxiety medications.

- Combining medications with behavioral methods such as desensitization is more effective than using medication alone.

Rage- and Anger-Motivated Behavior

Examples

- A stallion colt reared alone in a stall is vicious toward other horses. This is because he never had the opportunity to learn how to interact with other horses. He does not know that once he has become dominant he no longer has to keep fighting.

- A dominant dog bites its owner. In some cases the dog will be dominant over one family member and subordinate to others. Some examples of dominant behavior are growling when told to get off a couch or refusing to obey commands he has already learned such as "sit" or "stay."

- A dog guarding its owner bites the mailman or a veterinarian.

- A bull charges people in a pasture.

- A dominant dog bites a subordinate dog. Dogs are not democratic: feed and pet the dominant dog first to prevent attacks on subordinate dogs.

- A dog who has not been socialized to small children when he is a puppy bites a toddler. To prevent attacks on small children, puppies must be socialized to many different toddlers.

- A bull calf hand-reared away from other cattle attacks a person when he becomes mature. This occurs because he thinks he is a person and he views the person as a subordinate who needs to be dominated. Aggression toward people can be reduced by rearing young bulls and stallions in a social group with their own species.

- Animals fight over access to food.

- A horse or dog bites for no obvious reason. This is most likely to occur in low-fear animals who have been beaten into submission.
- A dog or horse who has been raised without contact with other animals gets into vicious fights with other animals.

Principles of Troubleshooting

- Rage- and anger-motivated behaviors are more likely to occur in confident, active, assertive animals and less likely to occur in flighty, shy, fearful animals. This applies to all species of animals.
- Rage- and anger-motivated behaviors are more likely to occur in animals who have not been socialized to other animals and people.
- In some cases, punishment might be appropriate especially with low-fear animals.
- Anger and rage are the motivators of a dominant animal fighting a subordinate animal over food or mates.
- Obedience training and teaching a puppy that people control its food helps prevent and control aggression toward people.
- Dog biting problems in adult dogs must be handled by a behavior professional.
- Aggression problems will often get worse if nothing is done to correct the behavior.
- Castration of grazing animals such as bulls and stallions at a young age will reduce aggressive behavior. Castration has less predictable effects on dogs, although it does reduce fighting between strange male cats.
- Dogs with a dominant personality need to be taught that if they want something they have to work for it. Example: make the dog sit before petting him or giving him a treat.
- A dog should be trained to obey every member of the family to prevent him from dominating one family member and being subordinate to another.

Predatory Chasing

Examples

- A dog chases cars or joggers.
- A dog chases a cat when the cat runs.
- A cat chases birds.
- A cat chases the red dot of a laser pointer around the house.
- Monticore, a pet tiger, attacked trainer Roy Horn when Horn fell down. The tiger's prey chase drive was triggered by sudden rapid movement.
- A dog attacks when a person tries to run away.

Principles of Troubleshooting

- Predatory aggression is totally different compared to other types of aggression. The circuits in the brain are separate.
- Predatory chasing is a hardwired instinctual behavior triggered by rapid movement. Dogs who chase cars or joggers are exhibiting predatory chase behavior.
- Stopping predatory chasing may require the use of a shock collar. I do not like shock collars, but this is one of the few situations where a harsh correction may be required. The dog should wear the shock collar for a few days before the first shock is administered so that he will not associate the shock with either the collar or his owner, but with the behavior.
- Young animals can be taught through socialization and learning what to chase and what not to chase. Socializing puppies to young children will help prevent dangerous prey drive behavior toward children. Remember: the chasing behavior is hardwired, but the sign stimulus that triggers it is learned.

Sociality-Motivated Behavior

Examples

- Cattle prefer to graze with the cattle they were raised with.
- Dogs run in packs.
- A dog cries when separated from people.
- Lambs raised by nanny goats will attempt to breed goats when they mature.
- A high-sociality Labrador stays off the furniture when she is praised for staying off the furniture.
- A puppy who is socialized at an early age with many different people and animals is more likely to become a friendly adult dog.
- A dog chews up the house when he is alone. Gradually teach the dog to tolerate longer periods alone or get another dog for a companion.
- High-sociality purebred Brahman cattle will seek more stroking from people than low-sociality Hereford cattle.
- A pair of geese stay "married" for life.
- Cats living in a barn are feral and stay away from people because they were not handled by people when they were kittens. Kittens who are socialized to people early in life will be friendlier toward people.

Principles of Troubleshooting

- Animals are motivated to seek companions.
- Animals prefer the company of the animals or people they were raised with.

- Both genetic factors and early rearing environment will affect intensity of sociality motivation.
- Puppies and kittens form social bonds during a critical period at the beginning of their lives. For puppies the critical period is the first twelve weeks of life; for kittens the critical period is the first seven weeks of life. Handle puppies and kittens with extra gentleness during this period. Failure to socialize young kittens and puppies often results in a fearful adult.
- Sociality has a genetic basis. Example: high sociality enables wolves to cooperate during hunting. Some species are more motivated by praise and companionship than others. You can train a dog using praise, but you must use food to train a cat.
- Cats are more trainable than people realize. *Clicker training* works well with cats. The reason people often believe dogs are trainable but cats are not is that dogs are tuned into people and want to please, so they do *incidental learning*. Example: young puppies will bound out of the car the instant the owner opens the door, which is dangerous. But by the time a dog is an adult, he will probably have stopped doing this and instead will remain inside the car and look at his owner expectantly. This learning has occurred naturally, or *incidentally*, over time as the owner has sometimes restrained the dog and on other occasions allowed the dog to get out of the car.
- A mother animal licking and caring for her babies is a sociality-motivated behavior.
- Labrador retrievers seek praise. Rewarding with praise is most effective in high-sociality dogs. There are individual differences in responsiveness to praise. The praise must be given within one second after the desired behavior occurs so the animal will make the correct associations.
- Praise alone may not be a sufficient motivator for training a low-sociality animal. Use a food reward.
- In high-sociality animals use praise and avoid punishment.
- Species with high-sociality motivation such as geese are more likely to pair-bond and mate year after year.
- In general, dogs have higher sociality motivation than cats. Food rewards work well for training cats. Some dogs will respond well to praise and stroking alone and other less-sociality-motivated dogs will respond best to praise combined with food rewards.

Pain-Motivated Behavior

Examples

- An arthritic dog reduces activity. Treating the arthritis will result in a more active dog.

- An animal limps after an injury.
- A dog hit by a car bites a person.
- An animal stands still or lies hunched up after surgery.
- A cat with a urinary tract problem eliminates outside the litter box. Medical problems are the cause of 30 percent of cat elimination problems.
- A dog stays away from the boundary of an invisible fence to avoid a shock.
- A dog bites after a child has repeatedly pulled its ear.

Principles of Troubleshooting

- Never punish pain-motivated behavior caused by a medical problem or an injury.
- Fear or rage motivation is sometimes confused with pain motivation. Pain-related aggression is most likely to occur in direct response to manipulation of a sore body part.
- Animals will avoid places or activities that are associated with painful stimuli.
- When an injured dog bites a person he is less likely to develop a biting problem compared to a dog motivated by aggression or fear to bite.
- Prey species animals such as cattle, sheep, and horses cover up pain-related behavior when people are watching. They do this in the wild to avoid being eaten be predators.
- Research shows that painkilling medications and local anesthetics are effective in animals and they should be used during and after surgery.
- Pain-related behaviors are likely to be most evident when an animal is watched with a video camera with no people around.

Novelty-Seeking Behavior

Examples

- A dog runs excitedly from room to room to smell all the new smells in a strange house.
- A horse approaches a flag on a pasture because he is attracted to both the flapping movement and the contrasting color that is different from that of the pasture.
- A pig roots vigorously in a fresh bale of straw or excitedly chews up a paper bag thrown in its pen.
- A horse points his ears toward a novel sound such as a beeping horn.
- A monkey in a laboratory presses a button many times each day to open a door so he can get a brief look outside his cage.
- Cattle in a pasture watch construction crews building a bridge.
- Brahman cattle nose a coat hanging on a fence, while Hereford cattle ignore it.

Novelty Avoidance Behavior (Fear-Motivated)

Examples

- A dog panics at a fireworks show.
- Cattle accustomed to cowboys on horseback panic when they first see a person on foot. The cattle perceive the person on foot as a novel new scary thing.
- A horse rears at a show when he sees balloons and flags.
- A cat panics when he sees a dog for the first time. His hair will stand on end and he will hiss and scratch.
- Cattle who were calm and tame at the home ranch ram fences and charge people at an auction.
- A horse who has not become habituated to bikes, balloons, and flags at home is more likely to get scared of these and rear at a horse show.
- An antelope at the zoo panics and crashes into the fence when she sees roofers on top of her barn. The roofers are perceived as novel, but people standing around the exhibit are tolerated because they are no longer novel.

Principles of Troubleshooting

- New novel things are most frightening when they are introduced suddenly, such as an umbrella opening in an animal's face.
- New things are attractive if the animal can voluntarily approach them. Allow a horse to approach and sniff a new saddle.
- The paradox of novelty: new things are both the most attractive and the most feared things to animals with flighty, nervous, high-strung genetics. An Arab horse is more likely to spook if a flag is suddenly waved in his face. But he would be more likely than a horse with calmer genetics to approach a flag placed in the middle of a large pasture.
- All animals should be gradually introduced to many new things and new places to prevent panic when they travel to a new place.
- New things must be introduced more slowly to nervous, high-strung animals than to calmer animals to avoid panic and fear. Some examples of new things that frighten animals are: a person on a horse's back, a trailer, a balloon, flags, bikes, a garage door opening quickly, or a costume for a horse show.
- High-strung, nervous animals are more aware of new things in their environment.
- Horses and cattle are likely to be afraid of novel things that have erratic, rapid movements such as flags and balloons. Dogs are likely to develop a fear of loud noise.
- Animal memories are specific. A horse perceives a person on his back and a person on the ground as two different things.

Hunger-Motivated Behavior

Examples

- An animal is trained to perform a new behavior to gain a food reward.
- Animals come in from the pasture at feeding time.
- Lionesses teach their cubs how to hunt and what to eat.
- A cat comes running when he hears the can opener opening his food.
- A dolphin learns to present his tail for blood sampling voluntarily in exchange for a food reward.

Principles of Troubleshooting

- In order for an animal to make the association between a food reward and a desired behavior the food must be given within one second after the desired behavior occurs.
- The advantage of either *target training* or *clicker training* is that it is much easier to use the correct timing for the reward. The animal associates the click of a handheld clicker with food. When target training is used, the animal learns to associate either touching or following a short stick with a ball on the end with a food reward. Many good books are available that give step-by-step instructions on how to clicker-train or target-train.
- Grazing animals prefer the forages they ate when they were young.
- Prey chasing and killing behavior is not always food-motivated. Young animals are taught what to hunt by their mothers. Both dogs and lions must learn that the things they kill are good to eat.

Examples of Sex-Motivated Behavior

Examples

- Normal mating behavior such as copulation.
- Intact male dogs congregate on the doorstep of the house where there is a female dog in estrus.
- A dog "humps" a person's leg.
- A male bird performs a mating display toward a person and refuses to mate with his own species. He will strut and fan out his tail feathers.
- A stallion is overly aggressive during mating and will suddenly rush and mount a mare without greeting her first.

Principles of Troubleshooting

- Neutering animals before puberty will prevent many sexually motivated behaviors later in life.
- Animals who are neutered when mature often keep some adult sexual

behaviors. For example: male cats neutered as adults continue to spray chairs and walls.

- Abnormal sexual behavior can be prevented by raising young animals in social groups with their own species.
- Animals will often attempt to breed with the individuals who raised them.
- Single-trait breeding for production or appearance traits such as large muscles can sometimes cause abnormal sexual behavior such as a rooster who injures hens during mating because the normal courting behavior has been bred out.
- Keeping young male animals in isolation may result in abnormal, overly aggressive mating behavior. Young animals need to learn social rules from adult animals of their own species.

Hardwired Instinctual Behavior, or Fixed Action Patterns

Examples

- The following behaviors listed in previous sections are hardwired: male cat spraying, normal copulation behaviors, prey chase drive, and bird mating dances.
- Species-specific dominance display behavior such as a broadside threat from a bull or dominance posture in dogs is a fixed action pattern. The dog has an erect posture with staring eyes, ears forward, and hackles raised on its back.
- Dogs have a natural instinct to avoid messing in their sleeping area. This is why putting a puppy in its crate helps to housebreak the puppy.
- A chick raised by people performs its mating dance to a person. The mating dance is instinctual but the sign stimulus that turns it on is learned.
- A dominant pig who bites people becomes submissive after a human shoves a board against the neck area where another dominant pig would bite. This approach is effective because it imitates the instinctual fighting behavior of the dominant pig. Slapping the pig on her hindquarters is not effective because it does not imitate the instinctual fighting behavior of a pig. *Exerting dominance does not mean beating an animal into submission. Exerting dominance means using the animal's natural method of communication.*
- Prey drive in dogs is triggered by the sign stimulus of rapid movement. This is why dogs chase cars and joggers.
- A dog performs the killing bite to a squirrel's neck.
- Nursing and sucking in infant animals are fixed action patterns. The sign stimulus that triggers nursing is an object placed in the infant's mouth. Calves will suck on a person's finger.

- A submissive dog rolls over in front of a dominant dog to stop him from attacking. The submissive dog voluntarily rolls over and is not pushed down by the dominant dog. When you are establishing dominance over a dog, teach him to roll over on his back using food and social rewards. Do not throw the dog down.
- A dog's bowing "let's play" posture is a fixed action pattern. The dog lowers his front end and his rear remains elevated.
- Egg retrieval behavior in geese is a fixed action pattern. A mother goose will retrieve any egg-sized object that rolls out of her nest. She will retrieve golf balls or cans along with her eggs.

Principles of Troubleshooting

- A fixed action pattern is hardwired into the brain and runs like a computer program.
- The fixed action pattern is turned on when it is released by a sign stimulus.
- Hormones will activate sexual fixed action patterns in mature animals.
- Other fixed action patterns, such as nursing or the play bow in dogs, are not influenced by hormonal cycles.
- In some species people can easily imitate aspects of the fixed action patterns to exert dominance over an animal. Raising a stick over your head to imitate the raised antler display of an elk is an example.

Behavior with Mixed Motivations

Examples

- Fear versus novelty seeking: Cattle approach a paper bag lying on the ground and jump back when the wind moves it.
- Sex versus fear: A dog smells a female in estrus and keeps approaching the female even when another dominant dog chases him off repeatedly. Sex starts the approach behavior and fear of the dominant dog makes him retreat when he gets too close.
- Fear versus instinctual maternal behavior: A young female dog is scared when she sees her first puppy but the fear disappears when the puppies begin nursing.
- Fear versus aggression: A mother defending her newborn babies may alternate between aggression and fear.

Principles of Troubleshooting

- In some cases, such as fear versus novelty and sex versus fear, the behavior may alternate between two or more conflicting motivators. In other cases, such as the new mother who initially fears her puppies but nurtures them once nursing has begun, the initial motivation is replaced by the competing motivation.

- Mixed motivations are sometimes hard to decipher. Making a list of the observed behaviors may help.

Environmentally Caused Abnormal Behavior

Examples

- Young puppies reared in barren kennels are more hyper and excitable compared to puppies raised with more social interaction with people.
- A parrot who lacks social companionship pulls out its feathers.
- Cribbing in horses (repetitively biting on a fence).
- Bar biting by sows in stalls where there is no straw or dirt available for rooting and chewing.
- Pacing in zoo animals kept in a small cage.
- Dog licks excessively and causes a sore on its paw. Often due to separation anxiety.
- Mice kept in barren wire cages will pace and circle. This most likely to occur when nobody is near them at night, because the activity of people around them attracts their attention. The abnormal behavior begins when there is little external stimulation from the environment.

Principles of Troubleshooting

- It is important to prevent these behaviors from starting because they are difficult to stop after they are established. Abnormal behaviors are most likely to occur in barren cages that contain no materials to manipulate, or when animals are raised in isolation.
- In barren environments, high-strung nervous animals are more likely to develop *stereotypies*—behaviors that the animal repeats over and over again—compared to calm, placid animals. Pacing, circling, bar biting, and cribbing are all examples of stereotypies.
- An animal's environmental needs depend on the species. Highly social animals such as dogs and horses need the companionship of other animals or people. Grazing animals such as horses and cattle need hay or grass. Burrowing animals such as rodents need materials to burrow and hide in. Animals who walk long distances such as polar bears and tigers need room to roam.
- The nervous system of young animals reared in barren kennels or laboratory cages may be damaged because the growing nervous system needs varied sensory input to develop normally.
- Some of the most abnormal behaviors that occur in barren environments are performed when the animal is undisturbed by people. When people enter the animals stop doing the abnormal behavior. Video cameras used for security systems are an inexpensive method for detecting abnormal behavior.

Genetically Caused Abnormal Behavior

Examples

- A dog suddenly bites for no reason due to *psychomotor epilepsy*. This condition first appeared in springer spaniels who were bred to have hyper-alert posture. The bite will come out of the blue and is not related to a particular person or place.
- A deaf, blue-eyed dog is hyper-excitable.
- A hyper-excitable high-producing egg layer hen beats her feathers off by flapping her wings against her cage.
- Roosters bred for large breasts sometimes kill hens during mating. They have lost their normal hardwired instinctual courtship behavior due to single trait breeding.
- Goats who have epileptic seizures and faint when they hear a loud noise.
- A hyper Dalmatian is difficult to train.
- Nervous pointer dogs who get frozen in the pointing posture. If pushed, they will tip over.

Principles of Troubleshooting

- Genetic defects are most likely to occur when animals are selected and bred for a single color, appearance, behavior, or productivity trait, such as rapid growth, blue eyes, a certain body shape, or a single behavior.
- These problems can be avoided by selecting breeding stock that are free of behavioral or structural defects such as poor leg conformation that can lead to lameness. Look at the *whole* animal.

TRAINING METHODS

Reward-Motivated Training (No Punishment)

Examples

- A dog fetches a newspaper and is rewarded with praise and petting.
- A dog learns basic, sit, heel, and stay commands. Reward with praise, petting, and a few treats.
- A drug-sniffing or rescue dog learns its task by receiving lots of praise and petting.
- A dolphin is trained to jump through hoops using food rewards.
- Animals are trained using food rewards to cooperate with veterinary procedures. Example: a dolphin presenting his tail to the vet for blood sampling.

- A dressage horse is taught a complex movement using clicker training. This form of training works well because a click sound that has been previously associated with food can be given within one second of the horse performing the desired movement.
- A rat in a laboratory behavior experiment learns that when the light is flashing he can press a lever and obtain food.

Principles

- Use no punishment. This means no stimuli that cause fear or pain.
- All operant and classical conditioning methods that use rewards are in this category. There are many books available on operant conditioning.
- Clicker or target training is effective for teaching new skills, tricks, or behaviors, especially in low-sociality animals where food is the best reinforcer.
- Standard rewards: praise, stroking, food, or a stimulus such as a click that has been associated with a food reward.
- Timing of the reward is critical so that the animal associates the reward with the desired behavior. Reward within one second after the desired behavior occurs.
- Ignore behaviors you wish to eliminate.
- You may also withdraw positive reinforcement to stop undesirable behavior. Withdrawing a reward is not the same thing as punishment.
- For dogs, praise is often the only reward needed. Cats and many other animals will require either food rewards or a stimulus associated with food such as a click because they have lower levels of sociality.
- Reward-motivated positive training methods are the best methods for teaching new skills, tricks, or behaviors. Trainers have personal preferences on training methods, but the important principle is to use positive reward-based methods.

Undesirable Behaviors That Occur Because a Person Has Inadvertently Rewarded Them and How to Correct Them

- A dog begs at the dinner table. Ignore the begging dog to stop the behavior.
- A horse paws the feed bucket before feeding. Wait until the horse stops pawing and then feed him.
- A horse pushes up against you. Instantly withhold treats or stroking until he stops pushing.
- A young puppy mouths your hand. Withdraw stroking and/or play and move your hands out of reach the instant a tooth is felt.

Punishment-Motivated Training

Examples

- A dog learns to stay in his yard after an electronic invisible fence has been installed. The animal learns that when he hears the tone coming from his collar he can avoid a shock by moving away from the boundary.
- Cattle stay away from an electric fence.
- A shock collar is used to stop a dog who is chasing cars, joggers, or deer. This is one of the few legitimate uses of a shock collar.
- A rat in a behavior experiment learns to avoid a shock by pressing a bar when a light comes on.

Principles of Troubleshooting

- Operant or classical conditioning in which a punishment such as a shock is used to stop an undesirable behavior. Example: a dog learns that he can avoid a shock by not chasing a jogger.
- Behaviors with a strong instinctual motivation such as deer chasing are least likely to respond to positive methods and more likely to respond to punishment.
- Beating an animal or other severe punishment to exert dominance over an animal is cruel and not very effective. Use obedience training or imitation of natural instinctual behavior to exert dominance.
- Do not use punishment to teach new skills or tricks. Reward-based methods work better and are the most humane.

MISUNDERSTOOD MOTIVATION

Several people have written to me to discuss situations where a person misunderstands a dog's motivation. The most common problem occurs when a person walking a dog screams at their dog when he lunges or barks at another dog. In some cases, but not all, the dog may misinterpret the screaming as "Help protect me against this other bad dog." This causes the dog to bark more instead of stopping. In these cases the person should talk quietly to the dog to tell him that the other dog is OK. Sometimes another command such as "sit" will stop the barking. The dog knows what "sit" means and when he sits will often stop the barking. It is also important to avoid screaming the dog's name at him. This may cause the dog to fear his own name. People working with animals must always keep asking themselves "How does the animal perceive this situation?"

Notes

Chapter 1: My Story

1. Burrhus Frederic Skinner, *Beyond Freedom and Dignity* (New York: Alfred A. Knopf, 1971).
2. You can see the cover online at http://www.goldbergcoins.net/catalogarchive/20010331/chap006.htm.
3. John J. Ratey, *A User's Guide to the Brain: Perception, Attention, and the Four Theaters of the Brain* (New York: Vintage Books, 2002); John J. Ratey and Catherine Johnson, *Shadow Syndromes: The Mild Forms of Major Disorders That Sabotage Us* (New York: Bantam, 1997).
4. O. I. Lovaas, "Behavioral Treatment and Normal Educational and Intellectual Functioning in Young Autistic Children," *Journal of Consulting and Clinical Psychology* 55 (1987): 3–9.
5. John Ross and Barbara McKinney, *Dog Talk: Training Your Dog Through a Canine Point of View* (New York: St. Martin's Press, 1995), pp. 71–72.
6. D. J. Simons and C. F. Chabris, "Gorillas In Our Midst: Sustained Inattentional Blindness for Dynamic Events," *Perception* 28 (1999): 1059–74.
7. Rita Carter's book, *Exploring Consciousness* (Berkeley, CA: University of California Press, 2002) has a photo on page 17.

Chapter 2: How Animals Perceive the World

1. N. J. Minshew and G. Goldstein, "Autism as a Disorder of Complex Information Processing," *Mental Retardation and Developmental Disabilities Research Reviews* 4 (1998):129–36.
2. C. J. Murphy, K. Zadnik, and M. J. Mannis, "Myopia and Refractive Error in Dogs," *Investigative Ophthalmology and Visual Science* 33 (1992): 2459–63.
3. Oliver W. Sacks, *An Anthropologist on Mars: Seven Paradoxical Tales* (New York: Vintage Books, 1996).

4. A Web site called "Pawsitive Training for Better Dogs" has some nice examples of color photographs as they would be seen by a dichromatic animal versus a trichromatic person. Dichromatic animals probably see a similar world to what people with color blindness see, but with much less saturated colors.

5. Arien Mack and Irvin Rock, *Inattentional Blindness: An Overview* (Cambridge: MIT Press, 1998). They did almost all their research with vision, but they have preliminary findings showing that people have inattentional blindness for touch and hearing, too.

6. Minshew and Goldstein, "Autism as a Disorder."

7. Paul D. MacLean, *The Triune Brain in Evolution: Role in Paleocerebral Functions* (New York: Kluwer Academic Publishers, 1990).

8. Elkhonon Goldberg, *The Executive Brain: Frontal Lobes and the Civilized Mind* (New York: Oxford University Press, 2002).

9. Rupert Sheldrake, *Dogs That Know When Their Owners Are Coming Home: And Other Unexplained Powers of Animals* (New York: Three Rivers Press, 2000).

10. Katy Payne, *Silent Thunder: In the Presence of Elephants* (New York: Penguin Books, 1999).

11. *National Geographic News,* July 8, 2002, http://news.national geographic.com/news/2002/07/0701_020702_elephantvibes.html.

12. Jianzhi Zhang and David M. Webb, "Evolutionary Deterioration of the Vomeronasal Pheromone Transduction Pathway in Catarrhine Primates," *Proceedings of the National Academy of Sciences* 100, no. 14 (July 8, 2003): 8337–41.

13. Oliver Sacks, *The Man Who Mistook His Wife for a Hat: And Other Clinical Tales* (New York: Touchstone, 1998).

14. Mack and Rock, *Inattentional Blindness,* pp. 176–77.

Chapter 3: Animal Feelings

1. L. Zecca, D. Tampellini, M. Gerlach, P. Riederer, R. G. Fariello, and D. Sulzer, "Substantia Nigra Neuromelanin: Structure, Synthesis, and Molecular Behaviour," *Journal of Clinical Pathology: Molecular Pathology* 54 (2001): 414–18.

2. D. Creel, "Inappropriate Use of Albino Animals as Models in Research," *Pharmacol Biochem Behav* 12, no. 6 (1980): 969.

3. Facts about albino Dobermans: http://www.geocities.com/~amazon doc/albinism/textframe4.html.

4. Brian Kilcommons and Michael Capuzzo, *Mutts: America's Dog* (New York: Warner Books, 1996), p. 13.

5. Pennisi, "Genetics: Genome Resources to Boost Canines' Role in Gene Hunts," *Science* 304 (2004): 1093–95.

6. Carlos Vila, Peter Savolainen, Jesus E. Maldonado, Isabel R. Amorim, John E. Rice, Rodney L. Honeycutt, Keith A. Crandall, Joakim Lundeberg, and Robert K. Wayne, "Multiple and Ancient Origins of the Domestic Dog," *Science* 276, no. 13 (June 1997): 1687–89.

7. D. Goodwin, J. W. S. Bradshaw, and S. M. Wickens, "Paedomorphosis Affects Visual Signals of Domestic Dogs," *Animal Behaviour* 53 (1997): 297–304.

8. Susan Milius, "The Social Lives of Snakes from Loner to Attentive Parent," *Science News* (March 27, 2004): 201.

9. National Institutes of Mental Health, "Teenage Brain: A Work in Progress," http://www.nimh.nih.gov/publicat/teenbrain.cfm.

10. Robert M. Joseph, "Neuropsychological Frameworks for Understanding Autism," *International Review of Psychiatry* 11 (July 8, 1999): 309–25.

11. Jaak Panksepp, *Affective Neuroscience: The Foundations of Human and Animal Emotions* (New York: Oxford University Press, 1998), pp. 27–28.

12. Ibid., p. 144.

13. Ibid., p. 291.

14. Ibid., p. 149.

15. Dr. Panksepp writes SEEKING in capital letters.

16. R. A. Fox and J. R. Millam, "Unpredictable Environments and Neophobia in Orange-Winged Amazon Parrots (*Amazona amazonica*)." Animal Behavior Society meeting, July 19–23, 2003, Boise, ID.

17. Panksepp, *Affective Neuroscience*, p. 161.

18. Joanna Burger, *The Parrot Who Owns Me: The Story of a Relationship* (New York: Random House, 2002).

19. Pat A. Wakefield and Larry Carrara, *A Moose for Jessica* (New York: Puffin, 1992).

20. Paul H. Hemsworth and G. J. Coleman, *Human Livestock Interactions: The Stockperson and the Productivity and Welfare of Intensively Farmed Animals* (New York: C.A.B. International, 1998).

21. Z. Wang, L. J. Young, G. J. De Vries, and T. R. Insel, "Voles and Vasopressin: A Review of Molecular, Cellular, and Behavioral Studies of Pair Bonding and Paternal Behaviors," *Prog Brain Res.* 119 (1998): 483–99.

22. J. T. Winslow and T. R. Insel, "Neuroendocrine Basis of Social Recognition," *Curr Opin Neurobiol* 14, no. 2 (April 2004): 248–53.

23. John M. Stribley and C. Sue Carter, *Proceedings of the National Academy of Science.* 1999 October 26; 96 (22): 12601–604, "Developmental Biology: Developmental Exposure to Vasopressin Increases Aggression in Adult Prairie Voles."

24. J. Panksepp, R. Meeker, and N. J. Bean, "The Neurochemical Control of Crying," *Pharmacol Biochem Behav* 12 (1980): 437–43.

25. J. Panksepp, P. Lensing, M. Leboyer, and M. P. Bouvard, "Naltrexone and Other Potential New Pharmacological Treatments of Autism," *Brain Dysfunction* 4 (1991): 281–300.

26. J. Panksepp, N. J. Bean, P. Bishop, T. Vilberg, and T. L. Sahley, "Opioid Blockade and Social Comfort in Chicks," *Pharmacol Biochem Behav* 13 (1980): 673–83.

27. J. A. Byers and C. B. Walker, "Refining the Motor Training Hypothesis for the Evolution of Play," *Am Nat* 146 (1995): 25–40.

28. PBS has a nice Web site with a simple three-dimensional "tour of the brain" that shows most of the areas I mention in this book, although it doesn't go into Paul MacLean's triune brain theory. But you can look up areas like the hypothalamus or the cerebellum and get a good picture of where they are in the brain along with a short summary of what they do. http://www.pbs.org/wnet/brain/3d/. Another excellent Web site that does cover the triune brain theory as well as all the parts of the brain I've mentioned in this book is run by an Oregon psychiatrist named Jim Phelps. http://www.psycheducation.org/emotion/triune%20brain.htm.

29. Rodolfo R. Llinas, *I of the Vortex: From Neurons to Self* (Cambridge, MA: MIT Press, 2002).

30. J. M. Faure and A. D. Mills, "Improving the Adaptability of Animals by Selection," in *Genetics and the Behavior of Domestic Animals,* ed. T. Grandin (San Diego: Academic Press, 1998), p. 235.

31. B. Knutson et al., "Selective Alteration of Personality and Social Behavior by Serotonergic Intervention," *Am J Psychiatry* 155 (1998): 373–79.

32. I want to be sure to add that not everyone agrees that Springer rage is related to epilepsy, so down the line researchers may develop a new explanation.

33. Susan Milius, "Beast Buddies: Do Animals Have Friends?" *Science News* 164, no. 18 (November 1, 2003): 282.

Chapter 4: Animal Aggression

1. Jeffrey J. Sacks et al., "Special Report: Breeds of Dogs Involved in

Fatal Human Attacks in the United States Between 1979 and 1998," *JAVMA* 217, no. 6 (September 15, 2000).

2. Ibid.

3. Credit for this goes to John Siegal and his work on cats at the University of Medicine in New Jersey.

4. Jaak Panksepp, *Affective Neuroscience: The Foundations of Human and Animal Emotions* (New York: Oxford University Press, 1998), p. 194.

5. Ibid., p. 198.

6. Debra Niehoff, *The Biology of Violence* (New York: Free Press, 2002).

7. Panksepp, *Affective Neuroscience,* p. 168.

8. Antonio R. Damasio, *Descartes' Error: Emotion, Reason, and the Human Brain* (New York: Grosset/Putnam, 1994).

9. Panksepp, *Affective Neuroscience,* p. 194.

10. M. J. Raleigh, M. T. McGuire, G. L. Brammer., D. B. Pollack, and A. Yuwiler, "Serotonergic Mechanisms Promote Dominance Acquisition in Adult Male Vervet Monkeys," *Brain Research* 559 (1991): 181–90.

11. Nicholas H. Dodman, *If Only They Could Speak: Stories about Pets and Their People* (New York: W. W. Norton, 2003), pp. 130–44.

12. The Monks of New Skete, *How to Be Your Dog's Best Friend: the Classic Training Manual for Dog Owners (Revised & Updated Edition)* (New York: Little, Brown, 2002).

13. Sacks et al., "Special Report."

14. Rachel Smolker, *To Touch a Wild Dolphin* (New York: Nan A. Talese, 2001).

15. William J. Broad, "Evidence Puts Dolphins in New Light, as Killers," *New York Times,* July 6, 1999, http://www.fishingnj.org/artdolph agress.htm.

16. "Pull Fido pull!" *New Scientist* 19 (January 2002): 24.

17. Michael D. Lemonick, "Young, Single and Out of Control," *Time* (October 20, 1997).

18. John McGlone, *Pig Production: Biological Principles and Applications* (Clifton Park, NY: Delmar Learning, 2002).

19. Ed Price and S. J. R. Wallach, "Physical Isolation of Herd Reared Hereford Bulls Increases Aggressiveness Towards Humans," *Appl Anim Behav Sci* 27 (1990): 263–67.

20. L. S. Shore, "The Question of Dogs, Off-Leash Safety, and Recreation," *A Review of the Literature on Dog Bites* (April 12, 2002): 2.

21. Nicholas Dodman, *The Dog Who Loved Too Much: Tales, Treatments, and the Psychology of Dogs* (New York: Bantam, 1996), p. 75.

22. Ibid., pp. 71–74.

23. Nicholas H. Dodman et al., "Comparison of Personality Inventories of Owners of Dogs With and Without Behavior Problems," *The International Journal of Applied Research* 2, no. 1.

24. P. M. Barrett, R. M. Rapee, M. M. Dadds, and S. M. Ryan, "Family Enhancement of Cognitive Style in Anxious and Aggressive Children," *Journal of Abnormal Child Psychology* 24 (1996): 187–99.

25. Z. Viryani, J. Topal, M. Gacsi, A. Miklosi, and V. Csanyi, "Dogs Respond Appropriately to Cues of Humans' Attemntional Focus," *Behav Process* 66, no. 2 (may 31, 2004): 161–72.

Chapter 5: Pain and Suffering

1. F. C. Colpaert et al., "Self-Administration of the Analgesic Suprofen in Arthritic Rats: Evidence of *Mycobacterium butyricum*-Induced Arthritis as an Experimental Model of Chronic Pain," *Life Sci* 27 (1980): 921–28.

2. A. V. Apkarian et al., "Prefrontal Cortical Hyperactivity in Patients with Sympathetically Mediated Chronic Pain," *Journal of Neuroscience Letters* 311, no. 3 (October 5, 2001): 193–97.

3. Leucotomies are starting to come back again as a treatment for severe pain.

4. Antoinio R. Damasio, *Descartes' Error: Emotion, Reason, and the Human Brain* (New York: Grosset/Putnam, 1994), p. 266.

5. Apkarian et al., "Prefrontal Cortical Hyperactivity"; M. R. Milad, I. Vidal-Gonzalez, and G. J. Quirk, "Electrical Stimulation of Medial Prefrontal Cortex Reduces Conditioned Fear in a Temporally Specific Manner," *Behav Neurosci* 118, no. 2 (April 2004): 389–94.

6. N. H. Kalin, S. E. Shelton, R. J. Davidson, and A. E. Kelley, "The Primate Amygdala Mediates Acute Fear but Not the Behavioral and Physiological Components of Anxious Temperament," *Journal of Neuroscience* 21, no. 6 (March 15, 2001): 2067–74.

7. Ibid.

8. Ruth A. Lanius et al., "The Nature of Traumatic Memories: A 4-T fMRI Functional Connectivity Analysis," *American Journal of Psychiatry* 161, no. 1 (January 2004): 36–44.

9. Randolph M. Nesse and George C. Williams, *Why We Get Sick: The New Science of Darwinian Medicine* (New York: Vintage, 1996).

10. Chong Chen et al., "Abnormal Fear Response and Aggressive Behavior in Mutant Mice Deficient for Calcium-Calmodulin Kinase II," *Science* 266 (October 14, 1994): 291–94.

11. Panksepp, *Affective Neuroscience,* p. 39.

12. Damasio, *Descartes' Error*, p. 41.
13. Ibid., p. 44.
14. David Allen, *Stress Free Productivity* (New York: Viking, 2001).
15. Jaak Panksepp has a good description of these studies in *Affective Neuroscience*, pp. 221–22.
16. I'd like to thank Jaak Panksepp for this example.
17. Panksepp, *Affective Neuroscience*, p. 221.
18. Ibid.
19. Ibid., p. 231.
20. S. Mineka, "Evolutionary Memories, Emotional Processing, and the Emotional Disorders," *The Psychology of Learning and Motivation* 28 (1992): 161–206; Arne Ohman and Susan Mineka, "The Malicious Serpent: Snakes as a Prototypical Stimulus for an Evolved Module of Fear," *Current Directions in Psychological Science* 12 (2003): 5–9.
21. Dr. de Waal writes, "The watching of skilled models firmly plants action sequences in the head that come in handy, sometimes much later, when the same tasks need to be carried out." Frans de Waal, *The Ape and the Sushi Master: Cultural Reflections of a Primatologist* (New York: Basic Books, 2001), p. 24.
22. Joseph LeDoux, *The Emotional Brain: The Mysterious Underpinnings of Emotional Life* (New York: Simon & Schuster, 1998).
23. If you're interested in this subject, I recommend Arthur Reber's *Implicit Learning and Tacit Knowledge*. Dr. Reber is one of the major researchers in the field. Arthur S. Reber, *Implicit Learning and Tacit Knowledge: An Essay on the Cognitive Unconscious (Oxford Psychology Series, No. 19)* (New York: Oxford University Press, 1996).
24. LeDoux, *Emotional Brain*, pp. 254–55.
25. Invisible fences come in an underground buried wire version, and an aboveground wireless version.
26. I've heard only one bad story about an invisible fence, but it's important so I want to share it. A vet told me he had a lady whose dog was getting more and more neurotic, whimpering and crying, refusing to go outside—the dog was frantic. It turned out either the collar or the main unit had malfunctioned, and the dog was getting shocked every time it went outdoors. I've never heard of that happening to any other dog and neither had the vet, so I'm telling you about it only because it's good to know it's possible. The vet felt terrible about it, because he didn't know the lady had an invisible fence, and he put the dog through all kinds of medical tests thinking something was medically wrong before they finally figured out the problem. Now he

asks all his dog owners if they're using an invisible fence just so he can have it in his records.

27. Laura Hillenbrand, *Seabiscuit: An American Legend* (New York: Ballantine, 2002).

28. Lyudmila N. Trut, "Early Canid Domestication: The Farm Fox Experiment," *American Scientist* 87, no. 2 (March-April 1999).

29. Benjamin Kilham and Ed Gray, *Among the Bears: Raising Orphaned Cubs in the Wild* (New York: Owl Books, 2003).

30. Derek Grzelewski, "Otterly Fascinating," *Smithsonian Magazine* (November 2002).

Chapter 6: How Animals Think

1. George Page, *Inside the Animal Mind: A Groundbreaking Exploration of Animal Intelligence* (New York: Broadway Books, 2001), pp. 77–78.

2. S. Watanabe, J. Sakamoto, and M. Wakita, "Pigeons' Discrimination of Paintings by Monet and Picasso," *Journal of the Experimental Analysis of Behavior* 63 (1995): 165–74.

3. J. M. Pearce, "The Acquisition of Concrete and Abstract Categories in Pigeons," in *Current Topics in Animal Learning,* L. Dachowski and C. F. Flaherty, eds. (Hillsdale, NJ: Lawrence Erlbaum, 1991), pp. 141–64.

4. Irene Pepperberg, *The Alex Studies: Cognitive and Communicative Abilities of Grey Parrots* (Cambridge, MA: Harvard University Press, 2000).

5. Autistic children have both poor expressive language and poor receptive language. It's not just that they can't talk, or can't talk very well; they can't understand other people when they talk, either.

6. Marian Stamp Dawkins, *Through Our Eyes Only? The Search for Animal Consciousness* (New York: Oxford University Press, 1998); Aubrey Manning and Marian Stamp Dawkins, *An Introduction to Animal Behaviour* (New York: Cambridge University Press, 1998).

7. Thomas Bugnyar, "Leading a Conspecific Away from Food in Ravens (*Corvus corax*)?" *Animal Cognition* Paper 1435–9456, 7, no. 2 (April 2004): 69–76.

8. Alex A. S. Weir, Jackie Chappell, and Alex Kacelnik, "Shaping of Hooks in New Caledonian Crows," *Science* 297, no. 5583 (August 9, 2002): 981.

9. Elizabeth Marshall Thomas, *The Hidden Life of Dogs* (New York: Pocket Books, 1996).

10. A controlled experiment means you have two groups of subjects, an *experimental group* and a *control group,* who are equivalent in every

way except for the *variable* being tested. A pharmaceutical study that gives the drug being tested to one group of subjects and a placebo to another group of subjects is a controlled experiment. The gold standard in experimental research is the *double-blind* study, where neither the subjects nor the experimenters know which group got which pill until after the study is over.

11. A. Bandura, "The Role of Imitation," in *Social Learning and Personality Development,* A. Bandura and R. H. Walters, eds. (New York: Holt, Rinehart & Winston, 1963); A. Bandura, *Social Modeling Theory* (Chicago: Aldine-Atherton, 1963).

12. Lynn Koegel and her co-author, Claire Lazebnik, a parent of an autistic child, have written an excellent book called *Overcoming Autism* that I recommend to all parents. Lynn Kern Koegel and Claire Lazebnik, *Overcoming Autism* (New York: Viking, 2004).

13. Thomas Nagel, "What Is It Like to Be a Bat?" *Philosophical Review* 83, no. 4 (October 1974): 435–50.

14. A few books have been written about Genie, but the question of how much language she acquired after her mother brought her to the authorities is in dispute. Up until the early 1990s it was believed that Genie learned words but no grammar or syntax. But a critical analysis of all the literature on Genie, done by Peter E. Jones at Sheffield Hallam University, found that she did learn grammar and syntax, and was still learning at the point when her mother refused to let scientists study her any longer. Peter E. Jones, "Contradictions and Unanswered Questions in the Genie Case," http://www.feralchildren.com/en/pager.php?df=jones1995.

15. Susan Schaller, *A Man Without Words* (Berkeley: University of California Press, 1995).

16. There are at least three different lines of evidence that religion is basic to the human brain: (1) religion is universal to all cultures, (2) identical twins separated at birth have the same degree of religiosity as adults, and (3) there is a "God part" of the brain in the temporal lobes that makes you feel the presence of God when it's stimulated. I suppose it's possible that the "God part" gets developed only through religious education, but if we're born with it, then it should be there for language-less people, too.

17. Jonathan Schooler at the University of Pittsburgh, personal communication.

18. K. Louie and M. A. Wilson, "Temporally Structured Replay of Awake Hippocampal Ensemble Activity During Rapid Eye Movement Sleep," *Neuron* 29, no. 1 (January 2001): 145–56.

19. Jeremy Gray, Washington University, quoted in Erica Goode, "Study Links Problem-Solving Skills to Brain 'g' Spot," *New York Times*, February 27, 2003.

20. Clive Thompson, "There's a Sucker Born in Every Medial Prefrontal Cortex," *New York Times* late edition, sec. 6, October 26, 2003.

21. George Miller published his famous paper, "The Magical Number Seven Plus or Minus Two," in 1956. The brain is built to hold on to five to nine separate bits of information at the same time. That's why phone numbers are seven digits. George A. Miller, "The Magical Number Seven, Plus or Minus Two: Some Limits on Our Capacity for Processing Information," *Psychological Review* 63 (1956): 81–97. Available online at: http://www.well.com/user/smalin/miller.html.

22. C. N. Slobodchikoff, "Cognition and Communication in Prairie Dogs," in *The Cognitive Animal: Empirical and Theoretical Perspectives on Animal Cognition*, Marc Bekoff, Colin Allen, and Gordon M. Burghardt, eds. (Cambridge, MA: MIT Press, 2002), pp. 257–64.

23. Ibid.

24. Sophie Yin and B. McCowan, "Barking in Domestic Dogs: Noise or Communication?" Animal Behavior Society meeting, July 13–17, 2002, Bloomington, IN.

25. Steven Pinker, *How the Mind Works* (New York: W. W. Norton, 1999).

26. B. Maess et al., "Musical Syntax Is Processed in Broca's Area: An MEG Study," *Nature Neuroscience* 4 (May 2001): 540.

27. Sandra Trehub, *Music and Infants* (New York: Psychology Press, 2005).

28. T. L. Gottfried, A. Staby, and D. Riester, "Relation of Pitch Glide Perception and Mandarin Tone Identification," http://www.lawrence.edu/fac/rewgottt/mandmusic.html.

29. P. M. Gray et al., "The Music of Nature and the Nature of Music" *Science* 291 (2001): 52–54.

29. Luis Baptista, quoted in S. Milius, "Music without Borders," *Science News* 157, no. 16 (April 15, 2000): 252.

31. Ibid.

32. Ibid.

33. Marc D. Hauser, Noam Chomsky, and W. Tecumseh Fitch, "The Faculty of Language: What Is It, Who Has It, and How Did It Evolve?" *Science* 298 (November 22, 2002): 1569–79.

34. Irene Pepperberg, "That Damn Bird: A Talk with Irene Pepperberg," http://www.edge.org/3rd_culture/pepperberg03/pepperberg_index.html.

Chapter 7: Animal Genius: Extreme Talents

1. K. Scalise, "Secret of the Squirrel Brain: Memory Tricks Investigated in New UC Berkeley Study." News release, 10/17/97, http://berkeley.edu/news/media/releases/97legacy/10_17_97a.html.

2. Dana Canedy, "Seizure-Alert Dogs May Get Seeing-Eye Status in Florida," *New York Times*, March 29, 2002.

3. C. Boesch and M. Tomasello, "Chimpanzee and Human Cultures," *Current Anthropology* 39, no. 5 (December 1998): 591–614, http://cogweb.ucla.edu/Abstracts/Boesch_Tomasello_98.html. Boesch and Tomasello write, "This may be called cumulative cultural evolution or the ratchet effect (by analogy with the device that keeps things in place while the user prepares to advance them further)."

4. Donna Williams, *Nobody Nowhere: The Extraordinary Autobiography of an Autistic* (New York: Random House, 1992).

5. Hyper-specificity and general intelligence don't go together in people or in domestic animals like horses and dogs. My guess is they probably don't go together in dolphins and birds, either, but I don't know.

6. A. Shah and U. Frith, "An Islet of Ability in Autistic Children: A Research Note," *Journal of Child Psychology and Psychiatry* 24 (1983): 613–20.

7. A. W. Snyder and D. J. Mitchell, "Is Integer Arithmetic Fundamental to Mental Processing?: The Mind's Secret Arithmetic," *Proceedings of the Royal Society*, London, B 266 (1999): 587–92.

8. Betty Edwards's book *Drawing on the Right Side of the Brain* starts students off with upside-down drawings. Douglas S. Fox, a writer who did an article about Dr. Snyder's work for *Discover* magazine, tried drawing the negative space around a pair of scissors and said, "I felt I was drawing individual lines, not an object, and my drawing wasn't half bad, either." He was drawing the separate parts of what he was looking at, not his unified concept. Betty Edwards, *The New Drawing on the Right Side of the Brain* (New York: Putnam, 1999). Douglas S. Fox, "The Inner Savant," *Discover,* vol. 23, no. 2 (February 2002), available online at http://centreforthemind.com/newsmedia/webarchive/discoverinnersavant.cfm.

9. A. W. Snyder et al., "Savant-like Skills Exposed in Normal People by Suppressing the Left Frontotemporal Lobe," *Journal of Integrative Neuroscience* 2, no. 2 (December 2003):149–58; B. L. Miller et al., "Emergence of Artistic Talent in Frontotemporal Dementia," *Neurology* 51, no. 4 (October 1998): 978–82; B. L. Miller and C. E.

Hou, "Portraits of Artists: Emergence of Visual Creativity in Dementia," *Arch Neurol* 61, no. 6 (June 2004): 842–44.

10. You can see the drawings if you do a Web search for "Savant-Like Skills Exposed in Normal People by Suppressing the Left Fronto-Temporal Lobe" and click on the pdf file. http://www.centreforthemind.com/whoweare/SavantSkillsJournal.pdf.

11. R. K. Wayne et al., "Multiple and Ancient Origins of the Domestic Dog," *Science* 276, no. 13 (June 1997): 1687–89.

12. Shelly Simonds, "Theory Suggests Greater Role for Man's Best Friend," *ANU Reporter* 29, no. 1. http://info.anu.edu.au/mac/Newsletters_and_Journals/ANU_Reporter/_pdf/vol_29_no_01/dogs.html.

13. Simon Benson, "Man and Canine a Top Team," *Daily Telegraph*, March 25, 2002.

14. I'm assuming it's smell rather than a subtle behavioral cue because I don't *think* a person would show tiny behavioral changes indicating that her blood sugar is getting dangerously low when she's asleep. But of course I don't know. If diabetics do have tiny behavioral changes when their blood sugar is low, a dog is going to be more likely to spot them than a person.

Selected Bibliography

Chapter 1: My Story

Grandin, T. 1995. *Thinking in Pictures*. New York: Vintage.

———. 1998. Handling Methods and Facilities to Reduce Stress on Cattle, *Veterinary Clinics of North America*, 14:325–41.

———. 2000. *Livestock Handling and Transport*. Wallingford, U.K.: CABI Publishing.

———. 2000. My Mind on a Web Browser: How People with Autism Think. *Cerebrum* (Winter): 13–22.

———. 2003 Transferring Results of Behavioral Research to Industry to Improve Animal Welfare on the Farm, Ranch, and the Slaughter Plant. *Applied Animal Behaviour Science* 81:215–28

Hov, C., et al. 2000. Artistic Savants. *Neuropsychiatry, Neuropsychology, and Behavioral Neurology* 13:29–38.

Voisinet, B. D., T. Grandin, J. D. Tatum, S. F. O'Conner, and J. J. Strothers. 1997. Feedlot Cattle with Calm Temperaments Have Higher Average Daily Gains Than Cattle with Excitable Temperaments. *Journal of Animal Science* 75:892–96.

Chapter 2: How Animals Perceive the World

Carter, Rita. 2002. *Exploring Consciousness*. Berkeley, CA: University of California Press.

Gladwell, Malcolm. 2001. Wrong Turn. *New Yorker,* June 11. Mr. Gladwell's articles can be found at http://www.gladwell.com/.

Grandin, T. 1996. Factors That Impede Animal Movement at Slaughter Plants. *Journal of the American Veterinary Medical Association* 209: 757–59.

Grandin, T., and M. J. Deesing. 1998. Behavioral Genetics and Animal Science. In *Genetics and the Behavior of Domestic Animals,* ed. T. Grandin. San Diego, CA: Academic Press: 1–31.

Heffner, R. S., and H. E. Heffner. 1983. Hearing in Large Animals: Horses (*Equus caballus* and Cattle C *Bos taurus*). *Behavioral Neuroscience* 97:299–311.

Jacobs, G. H., J. F. Deegan, and J. Netz. 1998. Photopigment Basis for Dichromatic Color Vision in Cows, Goats and Sheep. *Visual Neuroscience* 15:581–84.

Kleiner, K. 2004. What We Gave Up for Color Vision. *New Scientist* (January 2004): 12.

Krasnegor, Norman A., G. Reid Lyon, and Patricia S. Goldman-Rakic. 1997. *Development of the Prefrontal Cortex: Evolution, Neurobiology, and Behavior.* Baltimore: Paul H. Brookes Publishing.

Lanier, J. L., T. Grandin, R. D. Green, D. Avery, and K. McGee. 2000. The Relationship Between Reaction to Sudden Intermittent Movements and Sounds and Temperament. *Journal of Animal Science* 78: 1476–74.

Lemmon, W. B., and G. H. Patterson. 1964. Depth Perception in Sheep: Effects of Interrupting the Mother Neonate Bond. *Science* 145: 835–36.

McConnell, P. B. 1990. Acoustic Structure and Receiver Response in Domestic Dogs (*Canis familiarus*). *Animal Behavior* 39:897–904.

Miller, P. E., and C. J. Murphy. 1995. Vision in Dogs. *Journal of the American Veterinary Medical Association* 12:1623–34.

Revkin, A. C. 2001. Eavesdropping on Elephant Society. *New York Times.* January 9.

Talling, J. C., N. K. Waran, C. M. Wathes, and J. A. Lines. 1998. Sound Avoidance by Domestic Pigs Depends Upon Characteristics of the Signal. *Applied Animal Behaviour Science* 58, nos. 3–4:255–66.

Chapter 3: Animal Feelings

Craig, J. V., M. L. Jan, C. R. Polley, A. L. Bhagwat, and A. D. Dayton. 1975. Changes in Relative Aggressiveness and Social Dominance Associated with Selection for Early Egg Production in Chickens. *Poultry Science* 54:1647–58.

Craig, J. V., and J. C. Swanson. 1994. Review: Welfare Perspectives on Hens Kept for Egg Production. *Poultry Science* 73:921–38.

Danbury, T. C., C. A. Wecks, J. P. Chambers, A. E. Waterman-Pearson, and S. C. Kestin. 2000. Self Selection of the Analgesic Drug Carprofen by Lame Broiler Chickens. *Veterinary Record* 146:307–11.

Dugatkin, Lee Alan. 2004. *Principles of Animal Behavior.* New York: W. W. Norton.

Faure, J. M., and A. D. Mills. 1998. Improving the Adaptability of Animals by Selection. In *Genetics and the Behavior of Domestic Animals,* ed. T. Grandin, 235–64. San Diego, CA: Academic Press.

Goodwin, D., J. W. S. Bradshaw, and S. M. Wickens. 1997. Paedomorphosis Affects Visual Signals in Domestic Dogs. *Animal Behavior* 53:297–304.

Grandin, T. 2002. Do Animals and People with Autism Have True Consciousness? *Evolution and Cognition* 8:241–48.

Grandin, T., and M. J. Deesing. 1998. Genetics and Animal Welfare. In *Genetics and the Behavior of Domestic Animals,* ed. T. Grandin, 319–46. San Diego, CA: Academic Press.

Grandin T., M. J. Deesing, J. J. Struthers, and A. M. Swinker. 1995. Cattle with Hair Whorls Above the Eyes Are More Behaviorally Agitated During Restraint. *Applied Animal Behaviour Science* 46:117–23.

Grandin, T., T. N. Dodman, and L. Shuster. 1989. Effect of Naltrexone on Relaxation Induced by Lateral Flank Pressure in Pigs. *Pharmacological Biochemistry of Behavior* 33: 839–42.

Hemsworth, P. H., J. L. Barnett, and C. Hansen. 1981. The Influence of Handling by Humans on the Growth and Corticosteroids in the Juvenile Female Pig. *Hormones and Behavior* 15:396–403.

Hughes, D. P. 2004. *Songs of the Gorilla Nation.* New York: Random House.

Millman, S. T., I. J. Duncan, and T. M. Widowski. 2000. Male Broiler Fowl Display High Levels of Aggression Towards Females. *Poultry Science* 79:1233–41.

Webster, A. B., and J. E. Hurnik. 1991. Behavior, Production and Well Being in the Laying Hen. 2. Individual Production and Relationships of Behavior to Production and Physical Condition. *Poultry Science* 70:421–28.

Chapter 4: Animal Aggression

Beaver, B. V. 1999. *Canine Behavior: A Guide for Veterinarians.* Philadelphia: W. B. Saunders.

———. 2003. *Feline Behavior, A Guide for Veterinarians.* St. Louis: Saunders/Elsevier Science.

Gates, G. 2003. *A Dog in Hand.* Irving, TX: Tapestry Press.

Grandin, T., and J. Bruning. 1992. Boar Presence Reduces Fighting in Mixed Slaughter-Weight Pigs. *Applied Animal Behaviour Science* 33: 273–76.

Landsberg, G., W. Hunthausen, and L. Ackerman. 2003. *Handbook of*

Behavior Problems of the Dog and Cat. London: Saunders/Elsevier Science.

Niehoff, D. 1999. *The Biology of Violence*. New York: The Free Press, 54–114.

Price, E. O., and S. J. R. Wallach. 1990. Physical Isolation of Hand Reared Hereford Bulls Increases Their Aggressiveness Towards Humans. *Applied Animal Behavior Science* 27:263–67.

Smolker, R. 2002. *To Touch a Wild Dolphin*. New York: Anchor.

Turner, D. C., and P. Bateson. 2002. *The Domestic Cat: The Biology of Its Behavior* (2nd ed.). Cambridge: Cambridge University Press.

Chapter 5: Pain and Suffering

Apkarian, A. V., P. S. Thomas, B. R. Krauss, and N. M. Szevcrengi. 2001. Prefrontal Cortical Hyperactivity in Patients with Sympathetically Mediated Pain. *Journal of Neuroscience Letters* 311:193–97.

Bateson, P. 1991. Assessment of Pain in Animals. *Animal Behavior* 42:827–39.

Boissy, A. 1995. Fear and Fearfulness in Animals. *Quarterly Review of Biology* 70:165–91.

Colpaert, F. C., J. P. Taryre, M. Alliaga, and W. Koek. 2001. Opiate Self Administration as a Measure of Chronic Nociceptive Pain in Arthritic Rats. *Pain* 91: 33–34.

Freeman, W., and J. W. Watts. 1950. *Psychosurgery in the Treatment of Mental Disorders and Intractable Pain*. Springfield, IL: Charles C. Thomas Publisher.

Gentle, M. J., D. Waddington, L. N. Hunter, and R. B. Jones. 1990. Behavioral Evidence for Persistent Pain Following Partial Beak Amputation in Chickens. *Applied Animal Behavior Science* 27:149–57.

Grandin, T. 1997. Assessment of Stress During Handling and Transport. *Journal of Animal Science* 75:249–57.

Hansen, B. D., E. M. Hardic, and G. S. Carroll. 1997. Physiological Measurements After Ovariohysterectomy in Dogs. *Applied Animal Behavior Science* 51:101–9.

Rainville, P., G. H. Duncan, D. D Price, B. Carrier, and C. Bushnell. 1997. Pain Affect Encoded in Human Anterior Cingulate but Not Somatosensory Cortex. *Science* 277:968–71.

Rogan, M. T., and J. E. LeDoux. 1996. Emotion Systems, Cells Synaptic Plasticity. *Cell* 85:469–75.

Sheddon, L. V. 2003. The Evidence for Pain in Fish: The Use of Morphine as an Analgesic. *Applied Animal Behavior Science* 83:153–62.

Chapter 6: How Animals Think

Ackers, S. H., and C. N. Slobodchikoff. 1999. Communication of Stimulus Size and Shape in Alarm Calls of Gunnison's Prairie Dogs (*Cynomys gunnisoni*). *Ethology* 105:149–62.

Bekoff, Marc, Colin Allen, and Gordon M. Burghardt. 2002. *The Cognitive Animal: Empirical and Theoretical Perspectives on Animal Cognition*. Cambridge, MA: The MIT Press.

Dawkins, M. S. 1993. *Through Our Eyes Only: In the Search for Animal Consciousness*. New York: W. H. Freeman.

Derr, M. 2001. What Do Those Barks Mean? To Dogs, It's All Just Talk. *New York Times,* April 24, D6.

Domjan, Michael. 1998. *The Principles of Learning and Behavior,* 4th ed. New York: Brooks/Cole Publishing.

Grandin, T. 1998. Objective Scoring of Animal Handling and Stunning Practices in Slaughter Plants. *Journal of the American Veterinary Medical Association* 212:36–93.

———. 2001. Welfare of Cattle During Slaughter and the Prevention of Non-Ambulatory (Down) Cattle. *Journal of the American Veterinary Medical Association* 219:1377–82.

Gray, P. M., B. Kraus, J. Atema, R. Payne, C. Krumhansl, and L. Batista. 2001. The Music of Nature and the Nature of Music. *Science* 291:52–54.

Griffin, D. R. 2001. *Animal Minds: Beyond Cognition to Consciousness*. Chicago: University of Chicago Press.

Louie, K., and M. A. Wilson. 2001. Temporally Structured Replay of Awake Hippocampal Ensemble Activity During Rapid Eye Movement Sleep. *Neuron* 1:145–56.

Lund, Nick. 2002. *Animal Cognition*. New York: Taylor & Francis Group.

Lyon, B. E. 2003. Egg Recognition and Counting Reduce Costs of Avian Conspecific Brood Parasitism. *Nature* 422:495–98.

Milius, S. 2004. Where'd I Put That? Maybe It Takes a Bird Brain to Find the Car Keys. *Science News* 165:103–5.

Pearce, John M. 1997. *Animal Learning and Cognition: An Introduction,* 2nd ed. East Hove, UK: Psychology Press.

Pepperberg, I. M. 1999. *The Alex Studies: Cognitive and Communicative Abilities of Grey Parrots*. Cambridge, MA: Harvard University Press.

Rogers, Lesly J. 1997. *Minds of Their Own: Thinking and Awareness in Animals*. Boulder, CO: Westview Press.

Slobodchikoff, C. N., C. Kiriazis, C. Fischer, and E. Creef. 1991. Semantic Information Distinguishing Individual Predators in the Alarm Calls of Gunnison's Prairie Dogs. *Animal Behavior* 42:713–19.

Slobodchikoff, C. N. 2002. Cognition and Communication in Prairie Dogs. In M. Bekoff, C. Allen, and G. Burghardt, eds., *The Cognitive Animal.* Cambridge, MA: MIT Press.

Chapter 7: Animal Genius: Extreme Talents

Birbaumer, N. 1999. Rain Man's Revelations. *Nature* 399:211–12.

Judd, S. P. D., and T. S. Collett. 1998. Multiple Stored Views and Landmark Guidance in Ants. *Nature* 392:710–14.

Miller, B. L., K. Boone, J. L. Cummings, S. L. Read, and F. Mishkin. 2000. Functional Correlates of Musical and Visual Ability in Frontotemporal Dementia. *British Journal of Psychiatry* 176:458–63.

Miller, B. L., J. Cummings, F. Mishkin, K. Boone, F. Prince, M. Ponton, and C. Cotman. 1998. Emergence of Art Talent in Frontotemporal Dementia. *Neurology* 51:978–81.

Snyder, A. W., and J. D. Mitchell. 1999. Is Integer Arithmetic Fundamental to Arithmetic? The Mind's Secret Arithmetic. *Proceedings Royal Society* 266:587–92.

Snyder, A. W., E. Mulcathy, J. L. Taylor, D. J. Mitchell, P. Sachdew, and S. Gandevia. 2003. Savant Like Skills Exposed in Normal People by Suppressing the Left Fronto-Temporal Lobe. *Journal of Integrative Neurosciences* 2:149–58. All of Dr. Snyder's articles are available at the Web site of the Centre for the Mind, a joint venture of the Australian National University and the University of Sydney. http://www.centre forthemind.com/publications/publications.cfm.

Treffert, Darald A. 1989. *Extraordinary People: Understanding "Idiot Savants."* New York: HarperCollins. Dr. Treffert also maintains a useful Web page devoted to savant syndromes on the Web site of the Wisconsin Medical Society. http://www.wisconsinmedicalsociety.org/ savant/whatsnew.cfm.

Webner, R. 2001. Bird Navigation Computing Orthodromes. *Science* 291:264–65.

Acknowledgments

This book would not have been possible without the input from many people who shared their animal experiences with me. I am especially indebted to Mark Deesing and Jennifer Lanier. Mark has worked for me for over ten years on many different animal behavior projects, and Red Dog lives with him. Jennifer was my first Ph.D. student, and both she and Mark provided me with many insights during hours of discussions.

I also need to thank all the people in the livestock industry who helped me start my career. In the early seventies, Ted Gilbert at the Red River Feedyard in Arizona and Tom Rohrer, manager of the Swift plant, tolerated my autistic obsessions and recognized my ability. Other industry people who helped me establish my career were: Mike Chabot, Gary Oden, Rick Jordan, Frank Brocholi, Wilson Swilley, Raoul Baxter, Glen Moyer, and Jim Uhl.

Bonnie Buntain with the USDA was instrumental in commissioning me to do a USDA study that became the basis for the American Meat Institute scoring system for measuring animal welfare in slaughter plants. Janet Riley of AMI and the entire animal welfare committee have supported my work on the Institute's guidelines, which are now being used by major restaurant companies to audit animal welfare. Bob Langert at McDonald's Corporation and Darren Brown at Wendy's International had the vision to implement the use of the objective scoring system to greatly improve animal welfare in slaughter plants.

The Department of Animal Science at Colorado State University has been my academic home for the last fifteen years, and I am especially indebted to Bernard Rollin for recommending that I be hired.

Last, I need to thank Catherine Johnson, my co-author. She made this book possible. Her beautiful prose found my voice and let me tell my story. I also wish to thank Betsy Lerner, my agent, Susan Moldow, my publisher at Scribner, Beth Wareham, my editor, and Rica Allannic.

—Temple Grandin

People ask me what it was like to write a book with an autistic person.

The answer is, it was great. Temple has all the unsung virtues of people with autism. She is kind, even-tempered, and almost as focused as Lilly and Harley chasing their laser mouse. Temple loves animals and knows practically everything there is to know about them, and she is never "off-task."

In college Temple taught herself a skill she called *finding the basic principle*. She had to do this, because, as she was not a word person, the only way she could remember the details of her coursework was to corral them all inside one big principle. Today she can read through vast and conflicting research literatures on the brain, on memory, on emotion, on animal behavior and cognition—*Animals in Translation* draws on at least nine different fields—and *see* the One Principle that ties all these works together. As a result, *Animals in Translation* is more than a book of stories about Temple's life with animals. It is also a book about the nature of animals and autism and how they go together.

Temple, thank you. I will miss our talks.

Next on the list is Betsy Lerner.

In the beginning, Temple and I had a hard time trying to write just one book on two subjects. After we'd struggled through the fifth or sixth draft of a long and ever-expanding document and still hadn't produced a proper proposal, we were stuck. One professional told us we needed to start over from the beginning; another advised me just to drop the whole thing and move on.

That's when Temple called Betsy, who had edited Temple's book *Thinking in Pictures*, and was now working as an agent.

Betsy read the proposal, called, and said she knew what the problem was and how to fix it. She did, too. Temple and I performed minor surgery on what we'd written, and *voilà*. The proposal was done. Then Betsy whistled up an auction and the book found its home. It was magic.

Betsy, thank you.

It's to my own agent, Suzanne Gluck, that I owe thanks for this book happening at all, because it was Suzanne who put Temple and me together in the first place. This she did on the spur of the moment one day not long after I'd moved to New York. I wasn't looking for a project at the time, but Suzanne had heard from a colleague that Temple wanted to write a book about animals, and that was that. "You're the only one to write that book!" she said. I wouldn't have put it quite that forcefully myself, but I'm glad Suzanne did. She got the ball rolling directly after lunch, and now here we are.

Thank you!

I haven't seen too many authors thank their editor for being a hoot, so let me be the first: Beth Wareham, our editor at Scribner, is a big, brassy gal from Texas who gets animals, gets Temple, and immediately understood the book and what we were trying to do. She is a hoot! Like Betsy, she could *see* each chapter in a way Temple and I sometimes couldn't. Beth is the "third brain" on this book (though this is the place where I should say that any errors aren't hers, but mine and Temple's.)

Thank you, Beth.

I'm grateful to Susan Moldow, Publisher of Scribner, not only for her galvanizing enthusiasm and interest, but also for our book's title. If it wasn't easy for Temple and me to put animals and autism together in the same book, it was impossible to get them both inside a decent title. Susan did it—she found the basic principle! *Thank you.*

And thank you, Rica Allannic, also of Scribner, for your competence, your efficiency, and your ability to open attachments no matter how many Mac-to-Word and Word-to-Mac iterations they've been through.

I always cringe when I get to the part of an author's acknowledgments where he thanks his wife and children for cheerfully enduring outright abandonment in the years during which he labored over his manuscript. In my case the disappearing act lasted months, not years, but it was still too long. I'm grateful to my husband, Ed Berenson, to our children's caregiver, Martine Saidi, and to our children, Jimmy, Andrew, and Christopher, for hanging in there. In the midst of an extremely intense work year of his own, Ed took over virtually all homemaking chores on evenings and weekends, and Martine handled pretty much everything else during the week. The kids soldiered on. Thank you, everyone, for surviving.

I want to thank my friend and neighbor (and fellow dog lover) Laura Read, who is a clinical psychologist, for talking through with me and clarifying a number of the ideas in this book. And I'm especially grateful to my co-chair on the PTSA after-school program, Penny Muise, for doing all of the heavy lifting. I owe you, Penny.

Finally, I'm grateful to God or to the universe—to whoever is in charge of these things—for the blessings life has brought. Two of my children, Jimmy and Andrew, have autism. Their brother, Christopher, is what the professionals call typical. Having my children diagnosed with autism was certainly the worst thing that ever happened to me; there's no changing that. But Jimmy and Andrew and Christopher, and the father they adore, are the best thing that's happened, too.

It was my children who brought me to this book, and now this book has brought me back to my children. Today I see autistic people a little differently, and I see typical folks a little differently, too. As for our animal friends and acquaintances, I know I will never again look a dog or a horse or a cow or a parrot or any other bird or beast in the eye and see the same being I used to see.

I have Temple to thank for that, Temple and all the researchers, trainers, wranglers, handlers, and vets who've spent their lives listening to animals. Thanks to all.

—Catherine Johnson

Index

ABOUT THE AUTHORS

TEMPLE GRANDIN has redefined society's perception of what is possible for people with autism. Her world-famous "hug machine," a pressure device she invented to alleviate her own anxiety, led to the invention of pressure therapies for autistic people worldwide. She has been instrumental in explaining sensory sensitivity as well as how autistic people think. Grandin is perhaps best known, however, for being a passionate and effective animal advocate and for explaining to humans how animals think. She revolutionized animal movement systems and spearheaded reform of the quality of life and humaneness of death for farm animals. In fact, half the cattle in the United States and Canada are handled in systems she designed.

An associate professor at Colorado State University, Grandin holds a Ph.D. in animal science from the University of Illinois. She is the author of four books: *Thinking in Pictures: And Other Reports from My Life with Autism, Emergence: Labeled Autistic, Genetics and the Behavior of Domestic Animals,* and *Livestock Handling and Transport.* Through her company, Grandin Livestock Systems, she works with the country's fast food purveyors, including McDonald's, Wendy's, and Burger King, to monitor the conditions of animal facilities worldwide. She lectures widely on both animal science and autism and serves as a role model for hundreds of thousands of families and people with autism.

CATHERINE JOHNSON, Ph.D., is a writer specializing in the brain and neuropsychiatry. For seven years she served as a trustee of the National Alliance for Autism Research, returning to civilian life just in time to begin work with Temple Grandin on *Animals in Translation.* She is the mother of three boys, two of whom have autism, and lives with her husband and children in Irvington, New York.